SATURATED
MODEL
THEORY

Second Edition

Gerald E. Sacks
Harvard University, USA

SATURATED
MODEL
THEORY

Second Edition

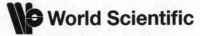

World Scientific

NEW JERSEY · LONDON · SINGAPORE · BEIJING · SHANGHAI · HONG KONG · TAIPEI · CHENNAI

Published by

World Scientific Publishing Co. Pte. Ltd.
5 Toh Tuck Link, Singapore 596224
USA office: 27 Warren Street, Suite 401-402, Hackensack, NJ 07601
UK office: 57 Shelton Street, Covent Garden, London WC2H 9HE

British Library Cataloguing-in-Publication Data
A catalogue record for this book is available from the British Library.

First edition published by W. A. Benjamin, Inc. (1972).

ISBN-13 978-981-283-381-5
ISBN-10 981-283-381-1

Typeset by Stallion Press
Email: enquiries@stallionpress.com

Printed in Singapore.

There are... three things that cause me to fear and that constitute for many writers a danger to their salvation or a loss of merit. These are: ignorance of the truth, misled or wanton statement of falsehood, and the haughty assertion of fact... I confess that I am at fault in all three respects.

The Metalogicon of
John of Salisbury

In Memory Of Professor Abraham Robinson

Contents

Section 0

Introduction

Introductory Remark 2009

*The intemperate young man who in 1972 wrote the Intro-
duction below has vanished beyond recall. His successor has made
numerous corrections, mostly typographical, some mathematical.
The sweeping assertions and dubious jokes remain. Altering
them would have been unjust to the author of long ago.*

This book was written to answer one question: "Does a
recursion theorist dare to write a book on model theory?" Consequently there are some observations scattered through it without
proof concerning the absoluteness (in the sense of Gödel [Gö1])
of model theoretic notions and the ordinals needed to define
them. For example Morley's notion of total transcendentality is
absolute, and the only ordinals needed to decide the total transcendentality of a theory T are the ordinals recursive in T. Part
of the blame belongs to B. Dreben who once asked with characteristic sweetness: "Does model theory have anything to do with
logic?" It is true that model theory bears a disheartening resem-
blance to set theory, a fascinating branch of mathematics with
little to say about fundamental logical questions. But the resem-
blance is more of manners than of ideas, because the central

1

notions of model theory are absolute, and absoluteness, unlike cardinality, is a logical concept. That is why model theory does not founder on that rock of undecidability, the generalized continuum hypothesis, and why the Łos conjecture is decidable: A theory T is κ-categorical if all models of T of cardinality κ are isomorphic. Łos conjectured and Morley proved (Theorem 37.4) that if a countable theory is κ-categorical for some uncountable κ, then it is κ-categorical for every uncountable κ. The property "T is κ-categorical for every uncountable κ" is of course an absolute property of T.

The notion of rank of 1-types was invented by Morley to prove Łos's conjecture. There are proofs of it that make no mention of rank, but they leave one ill prepared to prove Shelah's uniqueness theorem (Sec. 36). I have made rank a central idea of the book, because it is the central idea of current model theory. The assignment of rank to the 1-types realized by elements of structures makes it possible to prove theorems about structures by induction on rank. Not all 1-types associated with substructures of models of a theory T need have a rank; if they do, then T is said to be totally transcendental. Morley' s notion of rank was inspired by the Cantor–Bendixson differentiation of a closed subset of a compact Hausdorff space; however, the Morley derivative differs from the Cantor–Bendixson derivative in that the former commutes with the inverse limit operation. The Morley derivative is expounded in Sec. 29 as a transformation which acts on functors. Section 25 reviews the apparatus of category theory needed in Sec. 29.

· The title of this book is a misnomer. The coverage of saturated structures is far from complete: ultraproducts, a kind of canonical saturated extension of considerable importance, are not discussed. The title signifies a preference for the sort of model theory that minimizes syntactical questions. The book is briefer than it appears to be. The number of pages may be large, but the content of any one page is small because of the large size of type employed. A great deal of model theory has been left untouched, partly to achieve brevity, and partly to reflect the prejudices of the author. My mathematical taste favors new constructions and techniques, so I felt no urge to include important theorems whose proofs fail to be novel. Of course the limitations of my taste did not prevent me from repeating several of my favorite constructions.

It is no accident that the book suffers from a shortage of examples. As a rule examples are presented by authors in the hope of clarifying universal concepts, but all examples of the universal, since they must of necessity be particular and so partake of the individual, are misleading.

The least misleading example of a totally transcendental theory is the theory of differentially closed fields of characteristic 0 (DCF$_0$). Sections 40 and 41 are devoted to L. Blum's applications of Morley rank to DCF$_0$. There are many notable applications of model theory to algebra, and above all to theories of fields, but Blum was the first to apply something more than the compactness theorem (Corollary 7.2). (One of the most typical and influential uses of compactness in field theory is due to A. Robinson: Suppose F is a first order sentence (in the language of the theory of fields) that is true in every field of characteristic 0. Then there exists an integer n such that F is true in every field of characteristic $p \geq n$.) Blum showed every differential field of characteristic 0 has a prime differential closure. Her theorem follows from a general result of Morley (Theorem 32.4) which holds for all totally transcendental theories. An equally general result of Shelah implies the uniqueness of the prime differential closure (Theorem 41.4).

I am not a historian of model theory, so it is likely I have failed to assign credit justly to many who have contributed to the subject. Names have been attached parenthetically to most of the theorems, but it is morally certain that many of them were discovered independently by several persons (not all of whom were known to me), since it is rare that an inviting idea is the sole property of one mind. I hope no one will construe my ignorance as malice. It is not necessary to be a historian of model theory to realize that the subject owes its existence to the efforts of one man, Alfred Tarski.

Some precautions have been taken to make this book accessible to those with little logic. Definitions of "structures" and "sentences" are given in the early sections, and the commonsensical properties of "logical consequence" are sketched in enough detail in Sec. 7 to make the proof of the fundamental existence theorem (7.1) readable by all.

This book follows closely a course given at Yale University in the Fall of 1970, that course based on notes prepared by S. Simpson, those notes derived from a course given at the Massachusetts Institute of Technology in the Spring of 1969. A large debt — fortunately of the sort that never falls due — is owed to my students in both courses, who insisted relentlessly but rarely successfully that all proofs be complete and correct, to Jane MacIntyre who proved to be a very patient typist, and to my fellow model theorists, among them L. Blum, H. J. Keisler, G. Kreisel, A. H. Lachlan, M. Morley, A. Robinson, F. Rowbottom and S. Shelah, whose generous explanations opened me, contrary to my initial will, to a truly fascinating subject.

Cambridge, Massachusetts
March 15, 1972

Section 1

Ordinals and Diagrams

Ordinals are denoted by $\alpha, \beta, \gamma, \delta, \ldots$; each ordinal is equal to the set of all lesser ordinals; thus $\alpha = \{\beta | \beta < \alpha\}$. 0 is the empty set. Cardinals are denoted by $\kappa, \rho, \mu, \ldots$; a cardinal is an ordinal that cannot be put into one-to-one correspondence with any lesser ordinal. The infinite cardinals in increasing order are: $\omega_0 (= \omega), \omega_1, \omega_2, \ldots, \omega_\alpha, \ldots$; a set has cardinality κ if it can be put into one-to-one correspondence with κ. Card A is the cardinality of the set A. κ^+ is the least cardinal greater than κ. A set is countable if it is finite or has cardinality ω. A successor ordinal is an ordinal of the form $\alpha + 1$. λ denotes a limit ordinal, i.e. an ordinal neither 0 nor a successor. A cardinal κ is singular if there is a set $A \subseteq \kappa$ such that card $A < \kappa$ and κ is the least upper bound of A. A cardinal is regular if it is not singular.

An assertion that every diagram of the following sort

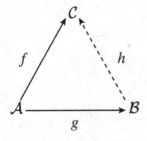

f and g have property P.
h has property Q.

5

can be completed as shown is to be understood as follows: For every f and g with property P, there exists an h with property Q such that $f = hg$.

An assertion that every diagram of the following sort

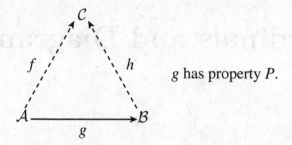

g has property P.

can be completed as shown is to be understood as follows: For every g with property P there exist f and h such that $f = hg$.

"iff" is an abreviation for "if and only if".

The end of a proof is signaled by □.

Section 2

Similarity Types of Structures

The coarsest classification of structures is by similarity type. Let N be the set of positive integers. A similarity type τ is a 5-tuple

$$\langle I, J, K, \theta, \Psi \rangle$$

such that $\theta: I \to N$ and $\Psi: J \to N$. A structure $\mathcal{A}$ of similarity type τ consists of:

(i) A nonempty set A called the universe of $\mathcal{A}$.

(ii) A family $\{R_i^{\mathcal{A}} \mid \epsilon I\}$ such that for each $i \in I$, $R_i^{\mathcal{A}}$ is a $\theta(i)$-place relation on A. An n-place relation on A is a subset of A^n, the Cartesian product of n copies of A. If $a_1, \ldots, a_n \in A$, then it is customary to write $R(a_1, \ldots, a_n)$ instead of $\langle a_1, \ldots, a_n \rangle \in R$.

(iii) A family $\{f_j^{\mathcal{A}} \mid j \in J\}$ such that for each $j \in J$, $f_j^{\mathcal{A}}$ is a $\Psi(j)$-place function on A. An n-place function on A is a function from A^n into A.

(iv) A subset $\{c_k^{\mathcal{A}} \mid k \in K\}$ of A called the set of distinguished elements of $\mathcal{A}$.

Structures are denoted by $\mathcal{A}, \mathcal{B}, \mathcal{C}, \ldots$ and the corresponding universes by $A', B, C', \ldots$; a useful form of notation for the structure a is

$$\langle A, R_i^{\mathcal{A}}, f_j^{\mathcal{A}}, c_k^{\mathcal{A}} \rangle_{i \in I, \ j \in J, \ k \in K}.$$

It can happen that $i \neq j$ and $R_i^{\mathcal{A}} = R_j^{\mathcal{A}}$; the same holds for the $f_j^{\mathcal{A}}$'s and the $c_k^{\mathcal{A}}$'s. The cardinality of $\mathcal{A}$ is defined to be the cardinality of A.

Consider the structure

$$\mathcal{A} = \langle A, +, \cdot, -, ^{-1}, 0, 1 \rangle,$$

where $+$ and $\cdot$ are 2-place functions on A, $-$ and $^{-1}$ are 1-place functions on A, and 0 and 1 are distinguished elements of A. The concept of field can be formulated so that every field has the same similarity types as $\mathcal{A}$, but $\mathcal{A}$ need not be a field, since the relations, functions and distinguished elements of $\mathcal{A}$ need not satisfy the axioms for fields.

Section 3

Monomorphisms and Substructures

A monomorphism m of $\mathcal{A}$ into $\mathcal{B}$, in symbols

$$m : \mathcal{A} \to \mathcal{B},$$

is a one–one map $m : \mathcal{A} \to \mathcal{B}$ ($a_1 = a_2$ iff $ma_1 = ma_2$) such that:

(i) $R_i^{\mathcal{A}}(a_1, \ldots, a_n)$ iff $R_i^{\mathcal{B}}(ma_1, \ldots, ma_n)$ for all $i \in I$.

(ii) $mf_j^{\mathcal{A}}(a_1, \ldots, a_n) = f_j^{\mathcal{B}}(ma_1, \ldots, ma_n)$ for all $j \in J$.

(iii) $mc_k^{\mathcal{A}} = c_k^{\mathcal{B}}$ for all $k \in K$.

It is assumed above that $\mathcal{A}$ and $\mathcal{B}$ are of the same similarity type. As a rule two structures mentioned in the same breath belong to the same similarity type.

A monomorphism between fields is nothing more nor less than a homomorphism.

$\mathcal{A}$ is a substructure of $\mathcal{B}$, in symbols

$$\mathcal{A} \subset \mathcal{B},$$

if $A \subset B$ and the inclusion map $i_A : A \subset B$ is a monomorphism. ($i_A a = a$ for each $a \in A$.) If $\mathcal{A} \subset \mathcal{B}$, then $\mathcal{B}$ is said to extend (or to be an extension of) $\mathcal{A}$. An isomorphism is a monomorphism that is onto. An isomorphism is indicated by $m : \mathcal{A} \xrightarrow{\approx} \mathcal{B}$ or by $\mathcal{A} \approx \mathcal{B}$. A monomorphism of $\mathcal{A}$ into $\mathcal{A}$ is called an endomorphism.

The "objects" of model theory are the structures. The "maps" of first order model theory are not the monomorphisms, which preserve merely the atomic structural properties, but rather the elementary monomorphisms, which preserve all first order properties. Let Q be the field of rationals and $\overline{Q}$ be the field of algebraic numbers. The inclusion map $i_Q: Q \to \overline{Q}$ is a monomorphism, but not an elementary one, since -1 has a square root in $\overline{Q}$ but not in Q. In Sec. 9, it will be shown that every monomorphism between algebraically closed fields is elementary.

Section 4

First Order Languages

Associated with each similarity type τ is a first order language $\mathcal{L}_\tau$. If $\mathcal{A}$ is a structure of type τ, then each sentence of $\mathcal{L}_\tau$ has a definite truth value in $\mathcal{A}$. The primitive symbols of $\mathcal{L}_\tau$ are:

(i) variables $x_1, x_2, x_3, \ldots$;
(ii) logical connectives $\sim$ (not), & (and), E (there exists), and $=$ (equals);
(iii) a $\theta(i)$-place relation symbol R_i for each $i \in I$;
(iv) a $\psi(j)$-place function symbol for each $j \in J$;
(v) an individual constant $\underline{c}_k$ for each $k \in K$.

The terms of $\mathcal{L}_\tau$ are generated by two rules:

(i) all variables and individual constants are terms;
(ii) if f_j is an n-place function symbol and $t_1, \ldots, t_n$ are terms, then $f_j(t_1, \ldots, t_n)$ is a term.

A constant term is one in which no variables occur.
The atomic formulas of $\mathcal{L}_\tau$ are given by:

(i) if t_1 and t_2 are terms, then $t_1 = t_2$ is an atomic formula called an equation;
(ii) if R_i is an n-place relation symbol and $t_1, \ldots, t_n$ are terms, then $R_i(t_1, \ldots, t_n)$ is an atomic formula.

11

The formulas of $\mathcal{L}_\tau$ are generated by two rules:

(i) every atomic formula is a formula;
(ii) if F and G are formulas, then $\sim F$, $F\&G$ and $(Ex_i)F$ are formulas. Read $\sim$ as 'not', $\&$ as 'and', and 'E' as 'there exists'.

The symbols $\vee$ (or), $\rightarrow$ (implies), $\leftrightarrow$ (if and only if), and (x_i) (for all x_i) are useful abbreviations:

(i) $F \vee G$ for $\sim (\sim F \& \sim G)$;
(ii) $F \rightarrow G$ for $(\sim F) \vee G$;
(iii) $F \leftrightarrow G$ for $(F \rightarrow G) \& (G \rightarrow F)$;
(iv) $(x_i)F$ for $\sim (Ex_i)\sim F$.

The notion of free variable of a formula is defined recursively. The recursion is on the number of steps needed to generate the formula:

(i)   if F is atomic and x_i occurs in F, then x_i is a free variable of F;
(ii) if x_i is a free variable of F and $j \neq i$, then x_i is a free variable of $(Ex_j)F$;
(iii) if x_i is a free variable of F, then x_i is a free variable of $\sim F$ and of $F\&G$.

In short the only way to kill a free variable x_i of F is to prefix F with (Ex_i).

A standard typographical convention concerning free variables is: all the free variables of $G(x_{i1}, \ldots, x_{in})$ lie among $x_{i1}, \ldots, x_{in}$. It is often convenient to drop subscripts and let $x, y, z, \ldots$ serve as variables.

A sentence is a formula without free variables.

A formula is said to have no quantifiers (or to be quantifierless) if it contains no expressions of the form (Ex) (there exists an x) or (y) (for all y).

The cardinality of $\mathcal{L}_\tau$ is the cardinality of the set of all formulas of $\mathcal{L}_\tau$. Clearly,

card $\mathcal{L}_\tau = \max(\omega,$ card $I,$ card $J,$ card $K)$.

$\mathcal{L}_\tau$ is often said to underlie $\mathcal{A}$.

Section 5

Elementary Equivalence

Suppose $\tau = \langle I, J, K, \theta, \psi \rangle$ is the similarity type of $\mathcal{A}$. Assume $K \cap A = 0$. Define τA to be the similarity type

$$\langle I, J, K \cup A, \theta, \psi \rangle.$$

The language $\mathcal{L}_{\tau A}$ is the language $\mathcal{L}_\tau$ augmented by a new individual constant $\underline{a}$ for each $a \in A$. Any structure $\mathcal{B}$ of type τ can be expanded to a structure $\langle \mathcal{B}, b_a \rangle_{a \in A}$ of type τA by choosing an element $b_a \in B$ for each $a \in A$. The map $\sigma \underline{a} = a$ has a unique extension to all constant terms of $\mathcal{L}_{\tau A}$:

(i) $\sigma \underline{a} = a$;

(ii) $\sigma \underline{c}_k = c_k^{\mathcal{A}}$;

(iii) $\sigma f_j(t_1, \ldots, t_n) = f_j^{\mathcal{A}}(\sigma t_1, \ldots, \sigma t_n)$, where t_i $(1 \leq i \leq n)$ is a constant term.

Let H be a sentence of $\mathcal{L}_{\tau A}$. The relation $\mathcal{A} \models H$ (read H is true in $\mathcal{A}$) is defined by induction on the number of steps needed to generate H.

(i) $\mathcal{A} \models t_1 = t_2$ iff $\sigma t_1 = \sigma t_2$.

(ii) $\mathcal{A} \models R_i(t_1, \ldots, t_n)$ iff $R_i^{\mathcal{A}}(\sigma t_1, \ldots, \sigma t_n)$.

(iii) $\mathcal{A} \models F \& G$ iff $\mathcal{A} \models F$ and $\mathcal{A} \models G$.

14

(iv) $\mathcal{A} \models \sim F$ iff it is not true that $\mathcal{A} \models F$.

(v) $\mathcal{A} \models (Ex_i)F(x_i)$ iff $\mathcal{A} \models F(\underline{a})$ for some $a \in A$.

If the sentence H is not true in $\mathcal{A}$, then it is said to be false in $\mathcal{A}$.

Let $F(x_1, \ldots, x_n)$ be a formula of $\mathcal{L}_\tau$, and let $a_1, \ldots, a_n \in A$. Then $\langle a_1, \ldots, a_n \rangle$ is said to satisfy (or realize) $F(x_1, \ldots, x_n)$ in $\mathcal{A}$ if $\mathcal{A} \models F(\underline{a}_1, \ldots, \underline{a}_n)$.

The universal closure of $F(x_1, \ldots, x_n)$ is $(x_1) \cdots (x_n)$ $F(x_1, \ldots, x_n)$. $F(x_1, \ldots, x_n)$ is said to be valid in $\mathcal{A}$ if its universal closure is true in $\mathcal{A}$.

$\mathcal{A}$ is elementarily equivalent to $\mathcal{B}$, in symbols

$$\mathcal{A} \equiv \mathcal{B},$$

means: $\mathcal{A} \models F$ iff $\mathcal{B} \models F$ for every sentence F of $\mathcal{L}_\tau$. It was Tarski's idea to classify structures up to elementary equivalence rather than isomorphism. It will be shown in Sec. 9 that any two algebraically closed fields of the same characteristic are elementarily equivalent. The notion of elementary equivalence is absolute in the sense of Gödel [Göl], while that of isomorphism is not.

Section 6

Elementary Monomorphisms

Let m be a map from A into B. m is an elementary monomorphism, in symbols

$$m : \mathcal{A} \overset{\equiv}{\longrightarrow} \mathcal{B},$$

means: $\mathcal{A} \models F(\underline{a}_1, \ldots, \underline{a}_n)$ iff $\mathcal{B} \models F(\underline{ma}_1, \ldots, \underline{ma}_n)$ for every formula $F(x_1, \ldots, x_n)$ of $\mathcal{L}_\tau$ and every sequence $a_1, \ldots, a_n \in A$. Observe that every elementary monomorphism is a monomorphism. (It will be shown in Sec. 9 that every monomorphism between algebraically closed fields of the same characteristic is elementary, and in Sec. 17 that every monomorphism between real closed fields is elementary.) Note that m is an elementary monomorphism iff

$$\langle \mathcal{A}, a \rangle_{a \in A} \equiv \langle \mathcal{B}, ma \rangle_{a \in A}.$$

Proposition 6.1. *Suppose $f : \mathcal{A} \to \mathcal{B}$ and $g : \mathcal{B} \to \mathcal{C}$. (i) If f and g are elementary, then gf is elementary. (ii) If g and gf are elementary, then f is elementary.*

Proof of (ii). Suppose $\mathcal{A} \models F(\underline{a}_1, \ldots, \underline{a}_n)$. Then $\mathcal{C} \models F(\underline{gfa}_1, \ldots, \underline{gfa}_n)$, since gf is elementary. But then $\mathcal{B} \models F(\underline{fa}_1, \ldots, \underline{fa}_n)$, since g is elementary. $\qquad \square$

$\mathcal{A}$ is an elementary substructure of $\mathcal{B}$, in symbols

$$\mathcal{A} \prec \mathcal{B}$$

if the inclusion map $i_A : A \subset B$ is an elementary monomorphism. If $\mathcal{A} \prec \mathcal{B}$, then $\mathcal{B}$ is said to be an elementary extension of $\mathcal{A}$. The extension is termed proper if $\mathcal{A} \neq \mathcal{B}$.

Section 7

The Fundamental Existence Theorem

Let S be a set of sentences of some language $\mathcal{L}_\tau$, and let F be a formula of $\mathcal{L}_\tau$. F is a logical consequence of S, in symbols

$$S \vdash F,$$

if F is among the formulas generated from S as follows:

(i) if $F \in S$, then $S \vdash F$;

(ii) if F is a logical axiom, then $S \vdash F$;

(iii) if $S \vdash F_i$ when $1 \le i \le n$, and if F is the result of applying some logical rule of inference to the sequence $F_1, \ldots, F_n$, then $S \vdash F$.

The axioms and rules of first order logic with equality can be found in any standard work on logic. They conform to common sense, so it should be clear that they possess all the properties cited below. If $S \vdash F$, then $S_0 \vdash F$ for some *finite* $S_0 \subset S$. The finitary character of the consequence relation $\vdash$ is exploited repeatedly in the proof of the fundamental existence Theorem 7.1.

S is consistent if no sentence of the form $F \& \sim F$ is a logical consequence of S. $\mathcal{A}$ is said to be a model of S, in symbols

$$\mathcal{A} \models S,$$

if every member of S is true in $\mathcal{A}$. If S has a model, then S is consistent, since every sentence which is a logical consequence of S is true in every model of S.

Theorem 7.1. *If S is a consistent set of sentences, then S has a model of cardinality $\leq \max(\omega, \text{card } S)$.*

Proof. [In the style of L. Henkin.] Let $\kappa = \max(\omega, \text{card } S)$. Suppose $\{\underline{c}_\delta | \delta < \kappa\}$ is a set of individual constants, none of which occur in the members of S. Let $\mathcal{L}$ be the language generated by the primitive symbols occurring in the members of S and the $\underline{c}_\delta$'s. Suppose $\{F_\delta(x) | \delta < \kappa\}$ is a list of all formulas of $\mathcal{L}$ whose only free variable is x. Choose a function $h : \kappa \to \kappa$ such that

(i) $\gamma < \delta$ implies $h\gamma < h\delta$;

(ii) $\gamma \leq \delta$ implies $\underline{c}_{h\delta}$ does not occur in $F_\gamma(x)$.

Define $S_\delta = S \cup \{(Ex)F_\gamma(x) \to F_\gamma(\underline{c}_{h\gamma}) | \gamma < \delta\}$.

Note that $\underline{c}_{h\delta}$ does not occur in S_δ. The consistency of S_δ for every $\delta < \kappa$ is established by transfinite induction. S_0 is consistent since $S_0 = S$. If λ is a limit ordinal and S_δ is consistent for every $\delta < \lambda$, then

$$S_\lambda = \cup\{S_\delta | \delta < \lambda\}$$

is consistent thanks to the finitary character of $\vdash$. Fix δ and suppose $S_{\delta+1}$ is not consistent with the intent of showing S_δ is not consistent:

$$S_{\delta+1} \vdash H\& \sim H$$
$$S_\delta \vdash [(Ex)F_\delta(x) \to F_\delta(\underline{c}_{h\delta})] \to [H\& \sim H]$$
$$S_\delta \vdash (Ex)F_\delta(x)\& \sim F_\delta(\underline{c}_{h\delta}).$$

since $\underline{c}_{h\delta}$ does not occur in S_δ, the part played by $\underline{c}_{h\delta}$ in generating logical consequences of S_δ, can be played just as well by some variable y not occurring in the generation. So

$$S_\delta \vdash (Ex)F_\delta(x)\& \sim F_\delta(y).$$

Since S_δ is a set of sentences, the universal closure of any consequence of S_δ is also a consequence of S_δ.

$$S_\delta \vdash (Ex)F_\delta(x) \& (y) \sim F_\delta(y).$$

Let $S_\kappa = \cup\{S_\delta | \delta < \kappa\}$. Then S_κ is consistent since every S_δ is consistent and κ is a limit ordinal. Let T be a maximal consistent set of sentences containing S_κ. The existence of T follows from Zorn's lemma and the finitary character of $\vdash$. Let F be an arbitrary sentence of $\mathcal{L}$. Since T is consistent, either $T \cup \{F\}$ or $T \cup \{\sim F\}$ is consistent. Since T is maximal, either $F \in T$ or $\sim F \in T$. Of course every consequence of T belongs to T.

A model $\mathcal{A}$ of S is constructed directly from T. For each individual constant c of $\mathcal{L}$, let

$$[c] = \{d | c = d \in T\}.$$

The universe A of $\mathcal{A}$ is $\{[c] | c \in \mathcal{L}\}$. The relations, functions, and distinguished elements of $\mathcal{A}$ are defined by:

(i) $R_i^{\mathcal{A}}([c_1], \ldots, [c_n])$ if $R_i(c_1, \ldots, c_n) \in T$;
(ii) $f_j^{\mathcal{A}}([c_1], \ldots, [c_n]) = [c]$ if $f_j(c_1, \ldots, c_n) = c \in T$;
(iii) $c_\kappa^{\mathcal{A}} = [c_\kappa]$.

In order to see that $\mathcal{A} \models S$, it is necessary to show — by induction on the number of steps needed to generate F — that $\mathcal{A} \models F$ iff $F \in T$, where F is any sentence of $\mathcal{L}$.

(i) Let t be a constant term of $\mathcal{L}$. For some $\delta < \kappa$, $F_\delta(x)$ is $t = x$. So $t = \underline{c}_{h\delta} \in T$. Thus for each constant term t_i of $\mathcal{L}$, there is an individual constant $\underline{c}_i$ of $\mathcal{L}$ such that $t_i = \underline{c}_i \in T$. Consequently

$$\mathcal{A} \models R_i(t_1, \ldots, t_n)$$
$$\text{iff } R_i^{\mathcal{A}}([c_1], \ldots, [c_n])$$
$$\text{iff } R_i(t_1, \ldots, t_n) \in T.$$

(ii) $\mathcal{A} \models \sim F$ iff $F \notin T$ iff $\sim F \in T$.

(iii) $\mathcal{A} \models F \& G$ iff $[\mathcal{A} \models F$ and $\mathcal{A} \models G]$ iff $F \& G \in T$.

(iv) Suppose $\mathcal{A} \models (Ex)F_\delta(x)$. Then $\mathcal{A} \models F_\delta(\underline{c})$

for some $[\underline{c}] \in A$; so $F_\delta(\underline{c}) \in T$ and $(Ex)F_\delta(x) \in T$. Now suppose $(Ex)F_\delta(x) \in T$. Then $F_\delta(\underline{c}_{h\delta}) \in T$. $\mathcal{A} \models F_\delta(\underline{c}_{h\delta})$ and $\mathcal{A} \models (Ex)F_\delta(x)$. $\qquad \square$

Theorem 7.1 is an amalgam of results due principally to K. Gödel, T. Skolem and A. Tarski. The proof of 7.1 follows a method originated by L. Henkin, a method that has become central to model theory. Perhaps this is so because Henkin's way of constructing a model takes into account the ultimate consequences of decisions made at intermediate stages of the construction.

A theory T is a consistent set of sentences. $T_1 \subset T_2$ means every logical consequence of T_1 is also a logical consequence of T_2. $T_1 = T_2$ means $T_1 \subset T_2$ and $T_2 \subset T_1$. T is said to be complete if either $T \vdash F$ or $T \vdash \sim F$ for every sentence F in the language of T. By 7.1 T is complete iff all models of T are elementarily equivalent. The complete theory of $\mathcal{A}$, denoted by $T\mathcal{A}$, is the set of all sentences of $\mathcal{L}_\tau$ true in $\mathcal{A}$, where τ is the similarity type of $\mathcal{A}$.

Corollary 7.2 (Compactness). Let S be a set of sentences such that every finite subset of S has an infinite model. Then S has a model of cardinality κ for every $\kappa \geq \max(\omega, \operatorname{card} S)$.

Proof. Let $\{\underline{c}_\delta | \delta < \kappa\}$ be a set of individual constants, none of which occur in the language of S. Let

$$W = S \cup \{\underline{c}_\delta \neq \underline{c}_\gamma | \delta < \gamma < \kappa\}.$$

If $V \subset W$ is finite, then V is consistent, since $V \cap S$ has an infinite model. By 7.1 W has a model $\mathcal{A}$ of cardinality $\leq \kappa$. But $\mathcal{A}$ must have cardinality at least κ, since the $\underline{c}_\delta$'s must name distinct elements of $\mathcal{A}$. $\qquad \square$

Corollary 7.3 (Upward Skolem–Löwenheim). Let $\mathcal{A}$ be an infinite structure of type τ. Then $\mathcal{A}$ has a proper elementary extension of cardinality κ for every $\kappa \geq \max(\text{card } \mathcal{A}, \text{card } \mathcal{L}_\tau)$.

Proof. Let T be the complete theory of $\langle \mathcal{A}, a \rangle_{a \in A}$. The models of T coincide with the elementary extensions of $\mathcal{A}$. Suppose $\underline{b}$ is an individual constant not occurring in T. Let

$$S = T \cup \{\underline{b} \neq \underline{a} \mid a \in A\}.$$

If $S_0 \subset S$ is finite, then $\mathcal{A}$ can be construed as a model of S_0 by letting $\underline{b}$ name some member of A not mentioned in S_0. By 7.2 S has a model $\mathcal{B}$ of card κ. Let $m : A \to B$ be the map defined by

$$ma = \underline{a}^{\mathcal{B}}.$$

Then $m : \mathcal{A} \to \mathcal{B}$ is elementary and $\underline{b}^{\mathcal{B}} \notin B - m[A]$.   □

Theorems 7.1, 7.2 and 7.3 suggest — only to those who leap to logical conclusions — that the construction of structures in the general setting afforded by model theory is limited to those structures that admit of simple description. If a structure with property P is desired, then the only hope seems to be: first find a theory T such that every model of T has property P, and then apply 7.1. But model theory, despite its immense generality, has methods more subtle than a direct appeal to 7.1. If this were not so, the subject would be dull indeed. One such method consists of iterated application of 7.3 to construct an elementary direct system followed by a use of 10.3.

The theory of fields (TF) consists of the following sentences:

(1) $(x)(y)(z)[(x + y) + z = x + (y + z)]$.
(2) $(x)[x + 0 = x]$.
(3) $(x)[x + (-x) = 0]$.
(4) $(x)(y)[x + y = y + x]$.
(5) $(x)(y)(z)[(x \cdot y) \cdot z = x \cdot (y \cdot z)]$.
(6) $(x)[x \cdot 1 = x]$.

(7) $(x)[x \neq 0 \rightarrow x \cdot x^{-1} = 1]$.

(8) $(x)(y)[x \cdot y = y \cdot x]$.

(9) $(x)(y)(z)[x \cdot (y + z) = (x \cdot y) + (x \cdot z)]$.

(10) $0 \neq 1$.

A model of TF is neither more nor less than a field.

Exercise 7.4 (A. Robinson). Let F be a sentence in the language of fields. Suppose F is true in every field of characteristic 0. Show there exists an integer n such that F is true in every field of characteristic greater than n.

Exercise 7.5. Suppose T is a theory without any infinite models. Show there exists an integer n such that every model of T has cardinality less than n.

Exercise 7.6 (R. Vaught). Suppose T is a countable theory with no finite models and κ is an infinite cardinal such that any two models of T of cardinality κ are isomorphic. Show T is complete.

Exercise 7.7. Show T is complete iff for every pair $\mathcal{A}, \mathcal{B}$ of models of T, there exist $\mathcal{C}$, $f : \mathcal{A} \xrightarrow{\equiv} \mathcal{C}$ and $g : \mathcal{B} \xrightarrow{\equiv} \mathcal{C}$.

Exercise 7.8. Devise elementarily equivalent linear orderings $\langle A, \leq \rangle$ and $\langle B, \leq \rangle$ such that $\langle A, \leq \rangle$ is a wellordering and $\langle B, \leq \rangle$ is not.

Exercise 7.9 (E. Artin). Let $\mathcal{A}$ be a field of characteristic 0, and $\mathcal{B}$ and $\mathcal{C}$ be algebraic extensions of $\mathcal{A}$. Suppose every polynomial in one variable with coefficients in $\mathcal{A}$ has a root in $\mathcal{B}$ iff it has a root in $\mathcal{C}$. Show $\mathcal{B}$ is isomorphic to $\mathcal{C}$ over $\mathcal{A}$. (J. Ax: the characteristic 0 assumption is not necessary.)

Section 8

Model Completeness

The notion of model completeness was inspired by Hilbert's Nullstellensatz (9.2). A theory T is model complete (A. Robinson) if every monomorphism between models of T is elementary. The diagram of $\mathcal{A}$, denoted by $D\mathcal{A}$, is the set of all atomic sentences and negations of atomic sentences true in $\langle \mathcal{A}, a \rangle_{a \in A}$. (If $\mathcal{A}$ is a multiplicative group, then $D\mathcal{A}$ conveys the same information as the multiplication table for $\mathcal{A}$.) The models of $D\mathcal{A}$ coincide with the extensions of $\mathcal{A}$. An existential formula is a formula of the form

$$(Ey_1) \cdots (Ey_m)H,$$

where $m \geq 0$ and H has no quantifiers.

Proposition 8.1. *If $G(x_1, \ldots, x_n)$ is an existential formula, $\mathcal{C} \models G(\underline{c}_1, \ldots, \underline{c}_n)$ and $g : \mathcal{C} \to \mathcal{D}$, then $\mathcal{D} \models G(\underline{gc}_1, \ldots, \underline{gc}_n)$.*

Proof. Let $G(x_1, \ldots, x_n)$ be

$$(Ey_1) \cdots (Ey_m)H(x_1, \ldots, x_n, y_1, \ldots, y_m),$$

where H has no quantifiers. Then

$$\mathcal{C} \models H(\underline{c}_1, \ldots, \underline{c}_n, \underline{d}_1, \ldots, \underline{d}_m)$$

24

for some $d_1, \ldots, d_m \in C$. But then

$$\mathcal{D} \models H(\underline{gc}_1, \ldots, \underline{gc}_n, \underline{gd}_1, \ldots, \underline{gd}_m),$$

and so $\mathcal{D} \models G(\underline{gc}_1, \ldots, \underline{gc}_n).$ $\qquad\qquad\square$

Theorem 8.2 (A. Robinson). *T is model complete iff (i) iff (ii).*

(i) $T \cup D\mathcal{A}$ is a complete theory for every model $\mathcal{A}$ of T.
(ii) For each formula F, there is an existential formula G such that $T \vdash F \leftrightarrow G$.

Proof. Suppose T is model complete and $\mathcal{A} \models T$. Let β_1 and β_2 be models of $T \cup D\mathcal{A}$ with the intent of showing $\beta_1 \equiv \beta_2$ in order to conclude from 7.1 that $T \cup D\mathcal{A}$ is complete. Since β_i $(i = 1, 2)$ is a model of $D\mathcal{A}$, there is a monomorphism $f_i : \mathcal{A} \to \beta_i$. f_i is elementary since T is model complete. Consequently

$$\beta_1 \models F(\underline{f_1 a}_1, \ldots \underline{f_1 a}_n)$$
$$\text{iff } \beta_2 \models F(\underline{f_2 a}_1, \ldots, \underline{f_2 a}_n).$$

Now suppose (ii) holds and $g : \mathcal{C} \to \mathcal{D}$ is a monomorphism between models of T. Let $F(x_1, \ldots, x_n)$ be a formula in the language of T. By (ii), there is an existential formula $G(x_1, \ldots, x_n)$ such that

$$T \vdash F(x_1, \ldots, x_n) \leftrightarrow G(x_1, \ldots, x_n).$$

Suppose $\mathcal{C} \models F(\underline{c}_1, \ldots, \underline{c}_n)$. Then $\mathcal{C} \models G(\underline{c}_1, \ldots, \underline{c}_n)$, $\mathcal{D} \models G(\underline{gc}_1, \ldots, \underline{gc}_n)$ by 8.1, and so $\mathcal{D} \models F(\underline{gc}_1, \ldots, \underline{gc}_n)$. Thus (ii) implies T is model complete.

Finally, suppose (i) holds in order to derive (ii). Let S be T augmented by the following sentences:

(1) $F(\underline{c})$, where $\underline{c}$ does not occur in T.
(2) $\sim K(\underline{c})$, where $K(x)$ is any existential formula such that $T \vdash K(x) \to F(x)$.

Assume for the sake of a reductio ad absurdum that S is consistent. By 7.1 S has a model $\mathcal{A}$. Clearly $\langle \mathcal{A}, a \rangle_{a \in A} \models F(\underline{c})$ for some $c \in A$. It follows from (i) that $T \cup D\mathcal{A} \vdash F(\underline{c})$. The finitary character of $\vdash$ implies

$$T \vdash Q(\underline{c}, \underline{a}_1, \ldots, \underline{a}_m) \to F(\underline{c}),$$

where $Q(\underline{c}, \underline{a}_1, \ldots, \underline{a}_m)$ is the conjunction of finitely many sentences of $D\mathcal{A}$. Since $\underline{c}, \underline{a}_1, \ldots, \underline{a}_m$ are individual constants not occurring in T, they can be replaced by variables. Thus

$$T \vdash Q(x, x_1, \ldots, x_m) \to F(x),$$

where Q has no quantifiers. So $T \vdash K(x) \to F(x)$ and $\mathcal{A} \models K(\underline{c})$, where $K(x)$ is the existential formula

$$(Ex_1) \cdots (Ex_m) Q(x, x_1 \ldots, x_m).$$

But the definition of S entails $\mathcal{A} \models \sim K(\underline{c})$.

The inconsistency of S means there must exist existential formulas $K_1(x), \ldots, K_n(x)$ such that

$$T \vdash K_i(x) \to F(x) \quad (1 \le i \le n)$$
$$T \vdash F(x) \to K_1(x) \vee \cdots \vee K_n(x).$$

Let $G(x)$ be $K_1(x) \vee \cdots \vee K_n(x)$. Then $T \vdash F(x) \leftrightarrow G(x)$, and $G(x)$ is logically equivalent to an existential formula.   □

It is possible to acquire an intuitive feeling for what happened in the last part of the proof of 8.2 by reflecting upon the following. A natural choice for G would be the "infinite disjunction" of all existential formulas that imply F.

Theorem 7.1 reduces the "infinite disjunction" to a finite disjunction. Such a reduction is termed a compactness phenomenon.

Exercise 8.3. Let T be a theory, F a sentence, and $\{G_i | i \in I\}$ a set of sentences. Suppose for each model $\mathcal{A}$ of T there is an

$i \in I$ such that $\mathcal{A} \models F \rightarrow G_i$. Show there exists a finite $J \subset I$ such that $T \vdash F \rightarrow \vee \{G_i | i \in J\}$.

Exercise 8.4. Suppose T is model complete and has a model imbeddable in every model of T. Show T is complete.

Section 9

Model Completeness of Algebraically Closed Fields

The theory of algebraically closed fields (ACF) extends the theory of fields (TF) by requiring that each nonconstant polynomial have a root. ACF is TF augmented by

$$(y_1) \cdots (y_n)(Ex)[x^n + y_1 x^{n-1} + \cdots + y_{n-1}x + y_n = 0]$$

fox each $n > 0$.

Theorem 9.1 (A. Robinson). ACF *is model complete.*

Proof. Let $f : \mathcal{A} \to \mathcal{B}$ be a monomorphism of algebraically closed fields. By 7.3 there exists $g : \mathcal{A} \overset{\equiv}{\to} \mathcal{A}_1$ and $h : \mathcal{B} \overset{\equiv}{\to} \mathcal{B}_1$ such that card $\mathcal{A}_1 = $ card $\mathcal{B}_1 > $ card $\mathcal{B}$. If there exists an isomorphism $k : \mathcal{A}_1 \overset{\approx}{\to} \mathcal{B}_1$ such that

$$kg = hf : \mathcal{A} \to \mathcal{B}_1,$$

then f is elementary by 6.1. It is safe to assume f, g and h are inclusion maps. Let U (respectively V) be a transcendence base for $\mathcal{A}_1$ (respectively $\mathcal{B}_1$) over $\mathcal{A}$. $\mathcal{B}$ is infinite, so $\mathcal{B}_1$ is uncountable; consequently card $U = $ card V. Let $k : U \to V$ be one-one and onto. Extend k to

$$k_1 : \mathcal{A}(U) \to \mathcal{A}(V)$$

28

so that k_1, is the identity on $\mathcal{A}$. k_1 can be extended to k_2 : $\mathcal{A}_1 \xrightarrow{\approx} \mathcal{B}_1$, since $\mathcal{A}_1$ (respectively $\mathcal{B}_1$) is the algebraic closure of $\mathcal{A}(U)$ (respectively $\mathcal{A}(V)$). □

The above argument turns on the rather special fact that any two uncountable, algebraically closed fields of the same cardinality and characteristic are isomorphic, so there is little chance of applying it in a general setting. In particular it cannot be used to prove the model completeness of the theory of real closed fields. That will be established in Sec. 17 by a virtually universal method involving saturated structures.

Corollary 9.2 (D. Hilbert). *Let S be a finite system of polynomial equations and inequations in several variables with coefficients in the field $\mathcal{A}$. If S has a solution in some field extending $\mathcal{A}$, then S has a solution in the algebraic closure of $\mathcal{A}$.*

Proof. let τ be the similarity type of $\mathcal{A}$. There is a sentence H in the language, $\mathcal{L}_{\tau A}$ such that for every field $\mathcal{B} \supset \mathcal{A}$: S has a solution in $\mathcal{B}$ iff $\mathcal{B} \models H$. By 9.1 every algebraically closed extension of $\mathcal{A}$ is an elementary extension of the algebraic closure of $\mathcal{A}$. □

A formula F is universal existential if it is of the form

$$(x_1) \cdots (x_m)(Ey_1) \cdots (Ey_n)G,$$

where $m \geq 0$, $n \geq 0$, and G has no quantifiers. A theory T is universal existential if there exists a theory W such that $T = W$ and every member of W is a universal existential sentence. ACF is a universal existential theory; that fact is no accident when viewed in the light of 9.1 and 9.3.

Proposition 9.3 (A. Robinson). *If T is model complete, then T is universal existential. (cf. Exercise 10.5.)*

Proof. A formula is said to be prenex normal if it is of the form

$$(Q_1 x_1) \cdots (Q_n x_n) H,$$

where H has no quantifiers, $n \geq 0$, and for each i $(1 \leq i \leq n)$, $(Q_i x_i)$ denotes either (Ex_i) or (x_i). The rank of a prenex normal formula is the number of alternations occurring in its quantifier prefix; an alternation consists of an occurrence of $(x)(Ey)$ or of $(Ex)(y)$. Thus the prefix

$$(x_1)(x_2)(Ex_3)(Ex_4)(x_5)(Ex_6)$$

has three alternations.

Let W be the set of all universal existential sentences provable in T. Every formula is logically equivalent to a prenex normal formula, so it suffices to show: if F is prenex normal and $T \vdash F$, then $W \vdash F$. The proof is by induction on the rank of F. Suppose $T \vdash F$

 (i) F is $(x_1) \cdots (x_n) G$, where G has lower rank than F. Then $T \vdash G$, $W \vdash G$, and $W \vdash F$.
 (ii) F is $(Ex_1) \cdots (Ex_n) G$, where G has lower rank than F. By 8.2, $T \vdash G \leftrightarrow K$ for some existential K. The formula $G \leftrightarrow K$ is logically equivalent to the conjunction of two prenex normal formulas, each of which has the same rank as G, so $W \vdash G \leftrightarrow K$, Then $T \vdash (Ex_1) \cdots (Ex_n) K$, $W \vdash (Ex_1) \cdots (Ex_n) K$, and $W \vdash F$.
(iii) F has rank 0. Then F is a logical consequence of some universal existential sentence belonging to W.   □

A formula is universal if it is of the form

$$(x_1) \cdots (x_n) H.$$

where $n \geq 0$ and H has no quantifiers. A theory T is universal if there exists a theory W such that $T = W$ and every member of W is a universal sentence.

Exercise 9.4 (Łos, Tarski). *Show T is universal iff every substructure of every model of T is a model of T.*

Exercise 9.5. Show that the theory of algebraically closed fields of characteristic 0 is complete. (A fine point: completeness requires a fixed rational value for 0^{-1}.)

Section 10

Direct Systems of Structures

The direct limit operation is needed to erect structures, e.g. saturated models of ω-stable theories, whose existence is not immediate from 7.1.

A directed set $\langle D, \leq \rangle$ consists of a set D with a partial ordering $\leq$ such that for any $i, j \in D$, there is a $k \in D$ with the property that $i \leq k$ and $j \leq k$: A direct system $\{\mathcal{A}_i, m_{ij}\}$ of structures and monomorphisms consists of a directed set $\langle D \leq \rangle$, a family $\{\mathcal{A}_i | i \in D\}$ of structures, and a family $\{m_{ij} : \mathcal{A}_i \to \mathcal{A}_j | i \leq j \in D\}$ of monomorphisms such that:

(a) $m_{ii} : \mathcal{A}_i \to \mathcal{A}_i$ is the identity,
(b) $m_{ik} = m_{jk} m_{ij}$ whenever $i \leq j \leq k$.

Let $A = \cup\{A_i \times \{i\} | i \in D\}$. If $\langle a, i \rangle, \langle b, j \rangle \in A$, then $\langle a, i \rangle \sim \langle b, j \rangle$ holds iff $m_{ik} a = m_{jk} b$ for some $k \in D$. Clearly $\sim$ is an equivalence relation. Let A_∞ be the set of all equivalence classes of A, and let $[a, i]$ be the equivalence class of $\langle a, i \rangle$.

The direct limit of $\{\mathcal{A}_i, m_{ij}\}$, denoted by $\varinjlim \mathcal{A}_i$ or $\mathcal{A}_\infty$, is a structure whose universe is A_∞. The relation

$$R^{A_\infty}([a_1, i_1], \ldots, [a_n, i_n])$$

holds iff

$$R^{A_k}(m_{i_1 k} a_1, \ldots, m_{i_n k} a_n)$$

holds for some k such that $i_t \leq k$ when $1 \leq t \leq n$. The functions and distinguished elements of $\mathcal{A}_\infty$ are defined similarly. The monomorphism

$$m_{i\infty} : \mathcal{A}_i \to \mathcal{A}_\infty$$

is defined by $m_{i\infty}a = [a, i]$: Clearly $m_{j\infty}m_{ij} = m_{i\infty}$ whenever $i \leq j$.

Theorem 10.1 (A. Tarski, R. Vaught). *If $\{\mathcal{A}_i, m_{ij}\}$ is a direct system of structures and elementary monomorphisms, then $m_{i\infty} : \mathcal{A}_i \to \mathcal{A}_\infty$ is elementary for all i.*

Proof. It suffices to show — by induction on the number of steps needed to generate $F(\underline{a}_1, \dots, \underline{a}_n)$ from the atomic formulas of $\mathcal{L}_{\tau A_i}$ — that for all i:

$$\mathcal{A}_i \models F(\underline{a}_1, \dots, \underline{a}_n) \text{ iff } \mathcal{A}_\infty \models F(\underline{m_{i\infty}a_1}, \dots, \underline{m_{i\infty}a_n}).$$

The sole induction step of interest starts with the assumption that $\mathcal{A}_\infty \models (Ex)G(x, \underline{m_{i\infty}a})$. Then

$$\mathcal{A}_\infty \models G(\underline{m_{j\infty}b}, \underline{m_{i\infty}a})$$

for some $j \in D$ and $b \in A_j$. Choose $k \in D$ so that $i \leq k$ and $j \leq k$. Then

$$\mathcal{A}_\infty \models G(\underline{m_{k\infty}m_{jk}b}, \underline{m_{k\infty}m_{ik}a}).$$

By induction $\mathcal{A}_k \models G(\underline{m_{jk}b}, \underline{m_{ik}a})$, so $\mathcal{A}_k \models (Ex)G(x, \underline{m_{ik}a})$. But then $\mathcal{A}_i \models (Ex)G(x, \underline{a})$, since m_{ik} is elementary. $\qquad\square$

The next proposition says that $\lim\limits_{\to} \mathcal{A}_i$ has the universal property associated with direct limits in the general setting of Sec. 25.

Proposition 10.2. *Let $\{\mathcal{A}_i, m_{ij}\}$ be a direct system of structures and monomorphisms. Suppose $\mathcal{B}$ is a structure and $\{f_i : \mathcal{A}_i \to \mathcal{B} \mid i \in D\}$ is a family of monomorphisms such that*

$f_j m_{ij} = f_i$ *whenever* $i \leq j$. *Then there exists a unique* $f : \mathcal{A}_\infty \to$ $\mathcal{B}$ *such that* $f m_{i\infty} = f_i$ *for all* i.

Proof. Define f by $f([a,i]) = f_i a$. Suppose $g : \mathcal{A}_\infty \to \mathcal{B}$ is such that $g m_{i\infty} = f_i$ for all i. Then $g([a,i]) = g(m_{i\infty}a) = f_i a$. $\qquad\square$

Let γ be an ordinal, and let $\{\mathcal{A}_\alpha | \alpha < \gamma\}$ be a family of structures such that $\mathcal{A}_\alpha \subset \mathcal{A}_\beta$ whenever $\alpha < \beta < \gamma$. $\{\mathcal{A}_\alpha | \alpha < \gamma\}$ is said to be a chain of length γ: $\{\mathcal{A}_\alpha | \alpha < \gamma\}$ can be construed as a direct system $\{\mathcal{A}_\alpha, i_{\alpha\beta}\}$, where $\alpha < \beta$ and $i_{\alpha\beta} : \mathcal{A}_\alpha \to \mathcal{A}_\beta$ is the inclusion monomorphism. The direct limit of $\{\mathcal{A}_\alpha, i_{\alpha\beta}\}$, call it $\mathcal{A}_\infty$, is easy to visualize, since A_∞ is nothing more than $\cup\{A_\alpha | \alpha < \gamma\}$, and a relation $R^{\mathcal{A}\infty}$ is merely $\cup\{R^{\mathcal{A}_\alpha} | \alpha < \gamma\}$. Thus it is customary to call $\mathcal{A}_\infty$ the union of $\{\mathcal{A}_\alpha | \alpha < \gamma\}$ and to write

$$\mathcal{A}_\infty = \cup\{\mathcal{A}_\alpha | \alpha < \gamma\}.$$

The chain $\{\mathcal{A}_\alpha | \alpha < \gamma\}$ is said to be elementary if $i_{\alpha\beta}$ is elementary whenever $\alpha < \beta$.

Corollary 10.3 (Elementary chain principle). *The union of an elementary chain is an elementary extension of every member of the chain.*

Proof. By 10.1. $\qquad\square$

Theorem 10.4. T *is model complete iff every diagram of the following sort*

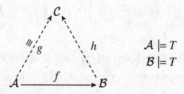

can be completed as shown.

Proof. If T is model complete, set $g = f$ and $h = i_B$. Suppose $f_0 : \mathcal{A}_0 \to \mathcal{B}_0$ is a monomorphism between models of T. Iterated use of the given diagram completion property of T makes it possible to erect the following infinite diagram.

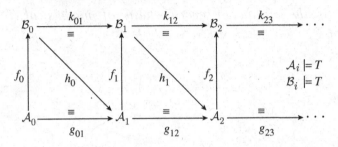

First h_0 and g_{01} are chosen, then f_1 and k_{01}, and so on.

Let $\mathcal{A}_\infty = \lim \mathcal{A}_i$ and $\mathcal{B}_\infty = \lim \mathcal{B}_i$. By 10.1 the monomorphisms $g_{i\infty} : \overrightarrow{\mathcal{A}_i} \to \mathcal{A}_\infty$ and $k_{i\infty} : \overrightarrow{\mathcal{B}_i} \to \mathcal{B}_\infty$ are elementary for all i. An isomorphism $f_\infty : \mathcal{A}_\infty \to \mathcal{B}_\infty$ is constructed so that $f_\infty g_{0\infty} = k_{0\infty} f_0$; it then follows from 6.1 that f_0 is elementary. Suppose $a \in A_\infty$. Choose i so that $g_{i\infty} a_i = a$ for some $a_i \in A_i$. Define $f_\infty a = k_{i\infty} f_i a_i$. To see that f_∞ is well defined, suppose $g_{i\infty} a_i = g_{j\infty} a_j = a$ and $i < j$. Then $g_{ij} a_i = a_j$ and

$$k_{i\infty} f_i a_i = k_{j\infty} k_{ij} f_i a_i = k_{j\infty} f_j g_{ij} a_i = k_{j\infty} f_j a_j.$$

Clearly f_∞ is a monomorphism. To see that f_∞ is onto, fix $b \in B_\infty$. Choose i so that $k_{i\infty} b_i = b$ for some $b_i \in B_i$. Let $a = g_{i+1,\infty} h_i b_i$. Then

$$f_\infty a = k_{i+1,\infty} f_{i+1} h_i b_i = k_{i+1,\infty} k_{i,i+1} b_i = b. \qquad \square$$

The criterion for model completeness of T supplied by 10.4 is difficult to apply because it requires the construction of an elementary monomorphism. If T is some theory of fields, then the available algebraic information about models of T is likely to say a great deal about monomorphisms and very little about elementary monomorphisms. The notion of saturated model will make it possible to give a criterion (Theorem 17.1) for model completeness similar to 10.4 but not requiring the construction of an elementary monomorphism.

Exercise 10.5 (Chang, Łos, Suszko). *T is universal existential iff the union of every chain of models of T is a model of T.*

Section 11

Skolemization of Structures

Skolemization is a device for factoring monomorphisms.

Let $\mathcal{L}$ be a language. Extend $\mathcal{L}$ to $\mathcal{L}^*$ by adding:

(i) a new individual constant $\underline{c}_F$ for each formula $F(x_1)$ of $\mathcal{L}$;

(ii) a new n-place function symbol f_F for each formula $F(x_1, \ldots, x_{n+1})$ of $\mathcal{L}$.

Define $\mathcal{L}^{\mathcal{S}}$, the Skolemization of $\mathcal{L}$, by: $\mathcal{L}^0 = \mathcal{L}$, $\mathcal{L}^{m+1} = (\mathcal{L}^m)^*$, and $\mathcal{L}^{\mathcal{S}} = \cup\{\mathcal{L}^m | m < w\}$.

Let T be a theory in the language $\mathcal{L}$. $T^{\mathcal{S}}$, the Skolemization of T, is a theory in the language $\mathcal{L}^{\mathcal{S}}$ obtained by adding the following Skolem axioms to T:

(i) $(Ex_1)F(x_1) \rightarrow F(\underline{c}_F)$;

(ii) the universal closure of

$$(Ex_{n+1})F(x_1, \ldots, x_n, x_{n+1}) \rightarrow F(x_1, \ldots, x_n, f_F(x_1, \ldots, x_n))$$

for every formula $F(x_1, \ldots, x_n, x_{n+1})(n > 0)$ of $\mathcal{L}^{\mathcal{S}}$.

A theory W admits elimination of quantifiers if for each formula F (in the language of W), there is a formula G without quantifiers such that

$$W \vdash F \leftrightarrow G.$$

Proposition 11.1. $T^{\mathcal{S}}$ admits elimination of quantifiers.

Proof. By induction on the number of steps needed to generate $F(x_1, \ldots, x_n)$ from the atomic formulas of $\mathcal{L}^{\mathcal{S}}$. Suppose $F(x_1, \ldots, x_n)$ is $(Ex_{n+1})H(x_1, \ldots, x_n, x_{n+1})$. Then

$$T^{\mathcal{S}} \vdash (Ex_{n+1})H(x_1, \ldots, x_n, x_{n+1})$$
$$\leftrightarrow H(x_2, \ldots, x_n, f_H(x_1, \ldots, x_n)).$$

By induction there is a quantifierless $G(x_1, \ldots, x_n)$ such that

$$T^{\mathcal{S}} \vdash H(x_1, \ldots, x_n, f_H(x_1, \ldots, x_n)) \leftrightarrow G(x_1, \ldots, x_n).$$

But then

$$T^{\mathcal{S}} \vdash F(x_1, \ldots, x_n) \leftrightarrow G(x_1, \ldots, x_n). \qquad \square$$

Let $\mathcal{A}$ be a model of T. $\mathcal{A}$ can be expanded to $\mathcal{A}^{\mathcal{S}}$, a model of $T^{\mathcal{S}}$, by choosing Skolem elements c_F and Skolem functions f_F to satisfy the Skolem axioms. $\mathcal{A}^{\mathcal{S}}$ is called a Skolemization of $\mathcal{A}$. In general the axiom of choice is needed to find a Skolemization of $\mathcal{A}$. If $\mathcal{D}$ is a model of $T^{\mathcal{S}}$, then $\mathcal{D}$ has a unique reduction to some model $\mathcal{A}$ of obtained by ignoring the Skolem elements and functions of $\mathcal{D}$. For each $X \subset A$, let $\mathcal{A}^{\mathcal{S}}(X)$, the Skolem hull of X in $\mathcal{A}^{\mathcal{S}}$, be the least substructure of $\mathcal{A}^{\mathcal{S}}$ whose universe contains X. The universe of $\mathcal{A}^{\mathcal{S}}(X)$ is generated by starting with X, adding the distinguished elements of $\mathcal{A}^{\mathcal{S}}$, and then closing under the functions of $\mathcal{A}^{\mathcal{S}}$.

Theorem 11.2 (Downward Skolem–Löwenheim). *For each $f : \mathcal{A} \to \mathcal{B}$ there exist $g : \mathcal{A} \to \mathcal{C}$ and $h : \mathcal{C} \overset{\equiv}{\to} \mathcal{B}$ such that $f = hg$ and card $\mathcal{C} \leq$ max (card $\mathcal{L}_\tau$, card $\mathcal{A}$), where τ is the similarity type of $\mathcal{A}$.*

Proof. Let $\mathcal{B}^{\mathcal{S}}$ be a Skolemization of $\mathcal{B}$, and let $\mathcal{C}$ be the reduction of $\mathcal{B}^{\mathcal{S}}(f[A])$ to a structure of type τ. By 11.1 $\mathcal{B}^{\mathcal{S}}(f[A]) \prec \mathcal{B}^{\mathcal{S}}$, because both $\mathcal{B}^{\mathcal{S}}$ and $\mathcal{B}^{\mathcal{S}}(f[A])$ are models of $T^{\mathcal{S}}$. Hence $\mathcal{C} \prec \mathcal{B}$. $\qquad \square$

Exercise 11.3 (T. Skolem). *Let F be a formula in the language of T such that $T^{\mathcal{S}} \vdash F$. Show $T \vdash F$.*

Section 12

Model Completions

A. Robinson's notion of model completion is useful for theorizing about theories of fields and for resolving questions concerning the solvability of systems of equations. It will be used in Sec. 40 to justify the definition of differentially closed field and to derive Seidenberg's Nullstellensatz for differential fields.

Let T and T_1 be theories in the same language. T_1 is a model completion of T if T_1 and T satisfy:

(i) if $\mathcal{A} \models T_1$, then $\mathcal{A} \models T$;
(ii) if $\mathcal{A} \models T$, then there exists a $\mathcal{B} \supset \mathcal{A}$ such that $\mathcal{B} \models T_1$;
(iii) if $\mathcal{A} \models T$, $\mathcal{A} \subset \mathcal{B}$, $\mathcal{A} \subset \mathcal{C}$, $\mathcal{B} \models T_1$ and $\mathcal{C} \models T_1$, then
$\langle \mathcal{B}, a \rangle_{a \in A} \equiv \langle \mathcal{C}, a \rangle_{a \in A}$.

If T_1 is a model completion of T, then T_1 is model complete. It can happen that T_1 and T satisfy (i) and (ii), T_1 is model complete but T_1 is not a model completion of T.

Theorem 12.1 (A. Robinson). *If T_1 and T_2 are model completions of T, then $T_1 = T_2$.*

Proof. Let $\mathcal{A}$ be an arbitrary model of T_1 in the hope of showing $\mathcal{A}$ is a model of T_2. A chain $\{\mathcal{A}_n | n < \omega\}$ of structures

39

is defined by:

(i) $\mathcal{A}_0 = \mathcal{A}$.

(ii) Assume $\mathcal{A}_{2n} \models T_1$; then $\mathcal{A}_{2n} \models T$ and so there exists an $\mathcal{A}_{2n+1} \supset \mathcal{A}_{2n}$ such that $\mathcal{A}_{2n+1} \models T_2$.

(iii) Assume $\mathcal{A}_{2n+1} \models T_2$; then $\mathcal{A}_{2n+1} \models T$ and so there exists an $\mathcal{A}_{2n+2} \supset \mathcal{A}_{2n+1}$ such that $\mathcal{A}_{2n+2} \models T_1$.

Let $\mathcal{A}_\infty = \cup\{\mathcal{A}_{2n}|n < \omega\} = \cup\{\mathcal{A}_{2n+1}|n < \omega\}$. $\{\mathcal{A}_{2n}|n < \omega\}$ is an elementary chain since T_1 is model complete. Hence $\mathcal{A}_0 \prec \mathcal{A}_\infty$ by the elementary chain principle (10.3). Similarly $\mathcal{A}_1 \prec \mathcal{A}_\infty$. So $\mathcal{A}_0 \prec \mathcal{A}_1$ by 6.1. Thus every model of T_1 is a model of T_2, and by symmetry every model of T_2 is a model of T_1. So $T_1 = T_2$ by 7.1 $\qquad\square$

Theorem 12.2 (A. Robinson). *The theory of algebraically closed fields is the model completion of the theory of fields.*

Proof. Suppose $\mathcal{A}$ is a field, and $\mathcal{B}$ and $\mathcal{C}$ are algebraically closed extensions of $\mathcal{A}$, with the intent of showing $\langle \mathcal{B}, a \rangle_{a \in A} \equiv \langle \mathcal{C}, a \rangle_{a \in A}$. Let $\mathcal{B}_1$ (respectively $\mathcal{C}_1$) be an algebraically closed, elementary extension of $\mathcal{B}$ (respectively $\mathcal{C}$) such that card $\mathcal{B}_1 =$ card $\mathcal{C}_1 >$ card $\mathcal{A}$. The existence of $\mathcal{B}_1$ and $\mathcal{C}_1$ follows from 7.3. Let U (respectively V) be a transcendence base for $\mathcal{B}_1$ (respectively $\mathcal{C}_1$) over $\mathcal{A}$. Then card $U =$ card $V =$ card $\mathcal{B}_1$. Let f be a one-one map of U onto V. Extend f to

$$f_1 : \mathcal{A}(U) \to \mathcal{A}(V)$$

so that f_1 is equal to the identity on $\mathcal{A}$. Then f_1 can be extended to $f_2 : \mathcal{B}_1 \xrightarrow{\approx} \mathcal{C}_1$, since $\mathcal{B}_1$ (respectively $\mathcal{C}_1$) is the algebraic closure of $\mathcal{A}(U)$ (respectively $\mathcal{A}(V)$). By 6.1 $\langle \mathcal{B}; a \rangle_{a \in A} \equiv \langle \mathcal{C}, a \rangle_{a \in A}$. $\qquad\square$

It will be shown in Sec. 17 that the theory of real closed fields is the model completion of the theory of ordered fields.

The proof will utilize an efficient, saturated structure criterion for identifying model completions.

The theory of linear ordering (LO) has three axioms:

a. $(x)(y)[x \le y \lor y \le x]$;
b. $(x)(y)(z)[x \le y \& y \le z \to x \le z]$;
c. $(x)(y)[x \le y \& y \le x \to x = y]$;

The theory of dense linear ordering without endpoints (DLO) is LO augmented to:

d. $(x)(y)(Ez)[x < y \to x < z < y]$;
e. $(x)(Ey)(Ez)[y < x < z]$.

($x < y$ is an abbreviation for $x \le y \& x \ne y$).

Theorem 12.3 (A. Robinson). *DLO is the model completion of LO.*

Proof. Any linear ordering $\mathcal{A}$ is readily extended to a dense linear ordering without endpoints by adding elements to $\mathcal{A}$ until all "gaps" are filled. So suppose $\mathcal{A} \models$ LO, $\mathcal{A} \subset \mathcal{B}$, $\mathcal{A} \subset \mathcal{C}$, $\mathcal{B} \models$ DLO and $\mathcal{C} \models$ DLO, but $\langle \mathcal{B}, a \rangle_{a \in A} \not\equiv \langle \mathcal{C}, a \rangle_{a \in A}$. Let $F(\underline{a}_0, \ldots, \underline{a}_n)$ be a sentence such that $\mathcal{B} \models F(\underline{a}_0, \ldots, \underline{a}_n)$ and $\mathcal{C} \models \sim F(\underline{a}_0, \ldots, \underline{a}_n)$. Let $\mathcal{A}_0$ be the finite substructure of $\mathcal{A}$ whose universe is $\{a_0, \ldots, a_n\}$. By 11.2 (downward Skolem–Löwenheim), there exist countable structures $\mathcal{B}_1$ and $\mathcal{C}_1$ such that $\mathcal{A}_0 \subset \mathcal{B}_1 \prec \mathcal{B}$ and $\mathcal{A} \subset \mathcal{C}_1 \prec \mathcal{C}$. Clearly

$$\langle \mathcal{B}_1, a_0, \ldots, a_n \rangle \not\equiv \langle \mathcal{C}_1, a_0, \ldots, a_n \rangle.$$

But this last is impossible because there exists an $f : \mathcal{B}_1 \overset{\approx}{\longrightarrow} \mathcal{C}_1$ such that $f a_i = a_i$ for all $i \le n$. f is constructed by means of a "back-and-forth" argument originated by Cantor. Let

$$\mathcal{B}_1 = \{b_i | i < \omega\} \quad \text{and} \quad \mathcal{C}_1 = \{\mathcal{C}_i | i < \omega\}$$

be enumerations with the property that $a_i = b_i = c_i$ for all $i \le n$. f is defined by induction on i.

Case 1. $i \leq n$. $fb_i = c_i$.

Case 2. $i > n$ and i is even. f has already been defined on some finite $B_0 \subset B_1$. Let b be that member of $B_1 - B_0$ whose subscript is least. b stands in a certain order relationship to the members of B_0. There must be a $c \in \mathcal{C}_1$ that stands in the same order relationship to the members of $f[B_0]$, because $\mathcal{C}$ is a dense linear ordering without endpoints and $f[B_0]$ is finite. Set $fb = c$.

Case 3. $i > n$ and i is odd. Same as case 2 with B_1 and C_1 interchanged. The range of f so far defined is some finite $C_0 \subset C_1$. Let c be that member of $C_1 - C_0$ whose subscript is least etc.   □

Section 13

Substructure Completeness

A theory T is substructure complete if $T \cup D\mathcal{A}$ is complete for every substructure $\mathcal{A}$ of a model of T.

Theorem 13.1. *T is substructure complete iff* (1) *iff* (2).

(1) *T admits elimination of quantifiers.*
(2) *Every diagram of the following sort*

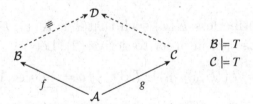

can be completed as shown.

Proof. Assume T is substructure complete in order to derive (1). Let $F(x)$ be a formula in the language of T, and let S be the following set of sentences:

(i) T, $F(\underline{c})$, where $\underline{c}$ does not occur in T.
(ii) $\sim K(\underline{c})$, where $K(x)$ is any quantifierless formula such that
   $T \vdash K(x) \rightarrow F(x)$.

43

Assume for the sake of a reductio ad absurdum that S is consistent. By 7.1 S has a model $\mathcal{A}$: Let $c \in A$ be the distinguished element named by $\underline{c}$, and let $\mathcal{C}$ be the least substructure of $\mathcal{A}$ having c as a member. Then

$$T \cup DC \vdash F(\underline{c}),$$

since $\mathcal{A} \models F(\underline{c})$ and $\mathcal{A}$ can be construed as a model of the complete theory $T \cup DC$. Each member of C is named by a constant term built up from $\underline{c}$, the individual constants of T and the functions of T. So

$$T \vdash K(\underline{c}) \rightarrow F(\underline{c}),$$

where $K(x)$ is quantifierless and $\mathcal{A} \models K(\underline{c})$. Since $\underline{c}$ does not occur in T,

$$T \vdash K(x) \rightarrow F(x).$$

But then $\mathcal{A} \models {\sim}K(\underline{c})$ by definition of S.

Since S is inconsistent,

$$T \vdash F(x) \rightarrow K(x)$$

for some quantifierless $K(x)$ such that $T \vdash K(x) \rightarrow F(x)$.

Now suppose (1) in order to derive (2). Let

$$W = T(\langle \mathcal{B}, b \rangle_{b \in B}) \cup DC \cup \{\underline{fa} = \underline{ga} | a \in A\}.$$

If $\mathcal{D} \models W$, then $\mathcal{D}$ completes the diagram as required. Suppose W is inconsistent. Then there exist typically $F(x_1, x_2)$, $G(x_1, x_2)$, $a \in A$, $b \in B - f[A]$ and $c \in C - g[A]$ such that:

(i) $\mathcal{B} \models F(\underline{b}, \underline{fa})$;
(ii) $G(x_1, x_2)$ is quantifierless and $\mathcal{C} \models G(\underline{c}, \underline{ga})$;
(iii) $F(\underline{b}, \underline{fa}) \,\&\, G(\underline{c}, \underline{ga}) \,\&\, \underline{fa} = \underline{ga}$ is inconsistent.

By (1) there is a quantifierless $H(x_2)$ such that

$$T \vdash (Ex_1)G(x_1, x_2) \leftrightarrow H(x_2).$$

Consequently $\mathcal{C} \models H(\underline{ga})$, $\mathcal{A} \models H(\underline{a})$ and $\mathcal{B} \models H(\underline{fa})$. But then

$$\mathcal{B} \models F(\underline{b}, \underline{fa}) \,\&\, (Ex_1)G(x_1, \underline{fa}),$$

an impossibility according to (iii).

Finally, assume (2) in order to show T is substructure complete. Let $f : \mathcal{A} \to \mathcal{B}_0$ and $g : \mathcal{A} \to \mathcal{C}_0$ be monomorphisms such that $\mathcal{B}_0$ and $\mathcal{C}_0$ are models of T. Iterated use of (2) makes it possible to erect to erect the following infinite diagram.

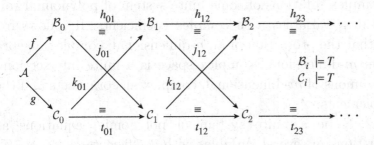

First h_{01} and k_{01} are chosen, then t_{01} and j_{01}, and so on. Let $\mathcal{B}_\infty = \lim \mathcal{B}_i$ and $\mathcal{C}_\infty = \lim \mathcal{C}_i$. By 10.1 the monomorphisms $h_{i\infty} : \mathcal{B}_i \overset{\rightarrow}{\to} \mathcal{B}_\infty$ and $t_{i\infty} : \mathcal{C}_1 \overset{\rightarrow}{\to} \mathcal{C}_\infty$ are elementary for all i. As in 10.4 an isomorphism

$$j_\infty : \mathcal{B}_\infty \overset{\approx}{\longrightarrow} \mathcal{C}_\infty$$

can be found so that $j_\infty h_{0\infty} f = t_{0\infty} g$. Then by 6.1

$$\langle \mathcal{B}_0, fa \rangle_{a \in A} \equiv \langle \mathcal{C}_0, ga \rangle_{a \in A}.$$

So $T \cup D\mathcal{A}$ is complete. $\qquad\square$

Theorem 13.2 (A. Robinson). *If T is a universal theory with a model completion T^*, then T^* admits elimination of quantifiers.*

Proof. By 13.1 it suffices to show T^* is substructure complete. Suppose $\mathcal{A}$ is a substructure of a model of T^*. Then every universal sentence provable in T^* must be true in $\mathcal{A}$. Consequently $\mathcal{A} \models T$, and so $T^* \cup D\mathcal{A}$ is complete. $\qquad\square$

Corollary 13.3 (A. Tarski, A. Robinson). *The theory of algebraically closed fields (ACF) admits elimination of quantifiers.*

Proof. By 12.2 and 13.2. □

Corollary 13.3 has some immediate consequences concerning algebraic sets and solvability of finite systems of polynomial equations. An n-dimensional, complex algebraic set is the set of all complex solutions of some finite system of polynomial equations in n variables with complex coefficients. It follows from 13.3 that the projection of an n-dimensional, complex algebraic set on m-dimensional complex space is a finite intersection of finite unions of m-dimensional, complex algebraic sets and their complements.

Let S be a finite system of polynomial equations and inequations in several variables with coefficients $c_1, \ldots, c_n$. The assertion that S has a solution can be expressed by some existential sentence $F(\underline{c}_1, \ldots, \underline{c}_n)$. By 13.3 there is a quantifierless formula $H(x_1, \ldots, x_n)$ such that

$$\mathrm{ACF} \vdash F(x_1, \ldots, x_n) \leftrightarrow H(x_1, \ldots, x_n).$$

Let $\mathcal{A}$ be any algebraically closed field containing $c_1, \ldots, c_n$. Then S has a solution in $\mathcal{A}$ iff

$$\mathcal{A} \models H(\underline{c}_1, \ldots, \underline{c}_n).$$

H provides an "algebraic" criterion for the solvability of S, since checking the truthvalue of $H(\underline{c}_1, \ldots, \underline{c}_n)$ in $\mathcal{A}$ amounts to nothing more than evaluating finitely many polynomials in $c_1, \ldots, c_n$ and noting which are zero and which are not.

Section 14

Countability Proviso with Exceptions

From now on every similarity type

$$\tau = \langle I, J, K, \theta, \Psi \rangle$$

will be countable; i.e. I, J and K will be countable sets. Thus every structure mentioned in every future section will have only countably many relations, functions and distinguished elements.

The only exceptions will be structures of the form $\langle \mathcal{A}, y \rangle_{y \in Y}$, where $\mathcal{A}$ is of countable similarity type and Y is an uncountable subset of A.

Every future theory T will be a set of sentences from some language $\mathcal{L}_\tau$, where τ is a countable similarity type. (If τ is countable, then $\mathcal{L}_\tau$ is countable.) The only exceptions will be of the form $T(\langle \mathcal{A}, y \rangle_{y \in Y})$, where Y is an uncountable subset of A.

The above proviso will simplify the statements of many results without sacrificing any of their essential content.

Section 15

Element Types

The notion of element type is needed to make a fine study of structures.

Let T be a complete theory. For each $n > 0$, let $F_n T$ be the set of all formulas in the language of T whose free variables lie among $x_1, \ldots, x_n$. Two formulas, F and G, of $F_n T$ are called equivalent if $T \vdash F \leftrightarrow G$. Let $[F]$ be the equivalence class of F. $B_n T$ is a Boolean algebra whose members are the $[F]'s$. The Boolean operations in B_n are defined by:

$$[F] \cup [G] = [F \vee G],$$
$$[F] \cap [G] = [F \, \& \, G],$$
$$c[F] = [\sim F].$$

A formula $F(x_1, \ldots, x_n)$ is said to be consistent with T if

$$T \vdash (Ex_1) \cdots (Ex_n) F(x_1, \ldots, x_n).$$

A set $S \subset F_n T$ is consistent with T if the conjunction of any finite number of members of S is consistent with T. An n-type p is a maximal consistent subset of $F_n T$. If each $F \in p$ is replaced by $[F]$, then the resulting subset of $B_n T$, ambiguously denoted by p, is a maximal dual ideal. Thus:

(i) if $F \in p$ and $G \in p$, then $F \& G \in p$;

48

(ii) $F \in p$ iff $\sim F \notin p$.

Every consistent subset of $F_n T$ can be extended to an n-type. $S_n T$ is the set of all n-types of T. ($S_n T$ is of course the Stone space of $B_n T$, the topological properties of $S_n T$ will be exploited in future sections.) Suppose $\mathcal{A} \models T$ and $a_1, \ldots, a_n \in \mathcal{A}$. $\langle a_1, \ldots, a_n \rangle$ is said to realize $p \in S_n T$ in $\mathcal{A}$ if

$$\mathcal{A} \models F(\underline{a}_1, \ldots, \underline{a}_n)$$

for every $F \in p$; i.e. $\langle a_1, \ldots, a_n \rangle$ satisfies every $F(x_1, \ldots, x_n) \in p$ in $\mathcal{A}$. If $\mathcal{B} \models T$ and $b, \ldots, b_n \in B$, then

$$\{F(x_1, \ldots, x_n) | \mathcal{B} \models F(\underline{b}_1, \ldots, \underline{b}_n)\}$$

is an n-type, namely the n-type realized by $\langle b_1, \ldots, b_n \rangle$ in $\mathcal{B}$.

Proposition 15.1. *Let $\mathcal{A}$ be an infinite structure, and let $Y \subset A$.*

(1) *If $p \in S_n T(\langle \mathcal{A}, y \rangle_{y \in Y})$, then there exists a $\mathcal{B} \succ \mathcal{A}$ such that p is realized in $\langle \mathcal{B}, y \rangle_{y \in Y}$ and card $\mathcal{B} = $ card $\mathcal{A}$.*

(2) *There exists a $\mathcal{B} \succ \mathcal{A}$ such that every $p \in S_n T(\langle \mathcal{A}, y \rangle_{y \in Y})$ is realized in $\mathcal{B}$ and card $\mathcal{B} \le $ card $\mathcal{A} \times 2^{\max(\omega, \text{card } Y)}$.*

Proof. To prove (1), let $\underline{c}_1, \ldots, \underline{c}_n$ be individual constants not mentioned in $T(\langle \mathcal{A}, y \rangle_{y \in Y})$. Let S be

$$T(\langle \mathcal{A}, a \rangle_{a \in A}) \cup \{F(\underline{c}_1, \ldots, \underline{c}_n) | F(x_1, \ldots, x_n) \in p\}.$$

$T(\langle \mathcal{A}, a \rangle_{y \in Y})$ can be regarded as an extension of $T(\langle \mathcal{A}, y \rangle_{y \in Y})$ if it is assumed that $\underline{a} = \underline{y}$ for all $y \in Y \subset A$. Then

$$(Ex_1) \cdots (Ex_n) F(x_1, \ldots, x_n) \in T(\langle \mathcal{A}, a \rangle_{a \in A})$$

for every $F(x_1, \ldots, x_n) \in p$. It follows that S is consistent. By 7.1 S has a model $\mathcal{B}$ such that card $\mathcal{B} = $ card $\mathcal{A}$, $\mathcal{A} \prec \mathcal{B}$ and p is realized in $\mathcal{B}$.

(2) follows from (1) and the elementary chain principle (10.3). Let $\kappa = $ card $S_n T(\langle \mathcal{A}, y \rangle_{y \in Y})$. By the countability proviso

(Sec. 14), $\kappa \leq 2^{\max(\omega,\,\mathrm{card}\,Y)}$. Let $\{p_\delta | \delta < \kappa\}$ be a well-ordering of $S_n T(\langle \mathcal{A}, y \rangle_{y \in Y})$. Define an elementary chain $\{\mathcal{A}_\delta | \delta \leq \kappa\}$ by transfinite recursion:

(i) $\mathcal{A}_0 = \mathcal{A}$.

(ii) Assume $\mathcal{A}_\delta$ is already defined so that $\mathcal{A}_0 \prec \mathcal{A}_\delta$. Then $\langle \mathcal{A}_0, y \rangle_{y \in Y} \equiv \langle \mathcal{A}_\delta, y \rangle_{y \in Y}$; and so p_δ, initially a member of $S_n T(\langle \mathcal{A}, y \rangle_{y \in Y})$, can now be regarded as a member of $S_n T(\langle \mathcal{A}_\delta, y \rangle_{y \in Y})$. By 15.1(1) there is an $\mathcal{A}_{\delta+1} \succ \mathcal{A}_\delta$ such that p_δ is realized in $\mathcal{A}_{\delta+1}$ and card $\mathcal{A}_{\delta+1} =$ card $\mathcal{A}_\delta$.

(iii) Assume that $\mathcal{A}_\delta$ is already defined for all δ less than some limit ordinal λ and that $\{\mathcal{A}_\delta | \delta < \lambda\}$ is an elementary chain. Let $\mathcal{A}_\lambda = \cup\{\mathcal{A}_\delta | \delta < \lambda\}$. Then $\mathcal{A}_\alpha \succ \mathcal{A}_\delta$ for all $\delta < \lambda$ by 10.3.

Let $\mathcal{B} = \mathcal{A}_\kappa$. Then p_δ is realized in $\mathcal{B}$, because $\mathcal{B} \succ \mathcal{A}_{\delta+1}$ and p_δ is realized in $\mathcal{A}_{\delta+1}$. It is readily checked by transfinite induction that card $\mathcal{A}_\delta \leq$ card $\mathcal{A} \times$ card δ for all $\delta \leq \kappa$. $\square$

Section 16

Saturated Structures

saturated structures are useful when devising model theoretic versions of syntactic notions, as in the characterization of model complete theories afforded by 17.1, and when studying categoricity, as in the characterization of ω-categoricity provided by 18.2.

Let $\mathcal{A}$ be an infinite structure, and let $Y \subset A$. $\mathcal{A}$ is saturated over Y if every $p \in S_1 T(\langle \mathcal{A}, y \rangle_{y \in Y})$ is realized in $\mathcal{A}$ (to be more precise, in $\langle \mathcal{A}, y \rangle_{y \in Y}$). $\mathcal{A}$ is saturated if $\mathcal{A}$ is saturated over every $Y \subset A$ such that card $Y <$ card A. Suppose κ is an infinite cardinal. $\mathcal{A}$ is κ-saturated if $\mathcal{A}$ is saturated over every $Y \subset A$ such that card $Y < \kappa$. The notion of saturation is not absolute in the sense of Gődel [Gől].

The traditional examples of saturated structures are the rationals as a dense linear ordering without endpoints, and the complex numbers as an algebraically closed field of characteristic 0.

Let $\mathcal{A} = \langle A, \leq \rangle$ be a linear ordering, i.e. a model of LO (Sec. 12). $\mathcal{A}$ is κ-dense if for each pair of sets $X, Y \subset A$ of cardinality $<\kappa$ such that

$$\mathcal{A} \models (x)(y)[x \in X \,\&\, y \in Y \to x < y],$$

it is the case that

$$\mathcal{A} \models (Ez)(x)(y)[x \in X \,\&\, y \in Y \to x < z < y].$$

$\mathcal{A}$ is dense iff $\mathcal{A}$ is ω-dense.

Proposition 16.1. *Let $\mathcal{A}$ be a dense linear ordering without endpoints. Then $\mathcal{A}$ is κ-saturated iff $\mathcal{A}$ is κ-dense.*

Proof. Let $\mathcal{A} = \langle A, \leq \rangle$ be a κ-dense linear ordering, and let $y \subset' A$ have cardinality $< \kappa$. Suppose $p \in S_1 T(\langle \mathcal{A}, y \rangle_{y \in Y})$. By 12.3 and 13.2 DLO admits elimination of quantifiers, so it is safe to assume that every $F(x_1) \in p$ is quantifierless. Consequently the formulas of p define the cut occupied by x_1 in the ordering of Y. Thus there exist A_1 and A_2 such that $Y = A_1 \cup A_2$ and p is equivalent to $\{\underline{a}_1 < x_1 \mid \underline{a}_1 \in A_1\} \cup \{x_1 < \underline{a}_2 \mid \underline{a}_2 \in A_2\}$. Then p must be realized in $\mathcal{A}$, since $\mathcal{A}$ is κ-dense.   □

Proposition 16.2. *Let $\mathcal{A}$ be an algebraically closed field. Then $\mathcal{A}$ is saturated iff $\mathcal{A}$ has infinite transcendence degree over its prime subfield.*

Proof. Let $\mathcal{A}$ be an algebraically closed and of infinite transcendence degree, and let $p \in S_1 T(\langle \mathcal{A}, y \rangle_{y \in Y})$, where card $Y <$ card A. By 15.1(1), p is realized by b in some $\mathcal{B} \succ \mathcal{A}$. If b is algebraic over Y, then $b \in A$. Suppose b is transcendental over Y. Let $\mathcal{C}$ be the least subfield of $\mathcal{A}$ containing Y. Choose $a \in A$ transcendental over Y; a exists because card $Y <$ card A and card A equals the transcendence degree of $\mathcal{A}$. Let $\overline{\mathcal{C}(a)}$ be the algebraic closure of $\mathcal{C}(a)$ in $\mathcal{A}$. Let

$$h : \overline{\mathcal{C}(a)} \to \mathcal{B}$$

be a monomorphism such that h is the identity on $\mathcal{C}$ and $ha = b$. By 9.1 h is elementary, so a realizes p in $\overline{\mathcal{C}(a)}$. But then a realizes p in $\mathcal{A}$, since the inclusion $\overline{\mathcal{C}(a)} \subset \mathcal{A}$ is elementary.   □

It follows from 16.2 that every uncountable, algebraically closed field is saturated. It will be shown in Sec. 37 that if T is a countable theory such that for some uncountable κ every model of T of cardinality κ is saturated, then every uncountable model of T is saturated.

Theorem 16.3 (M. Morley, R. Vaught). *If $\mathcal{A}$ and $\mathcal{B}$ are saturated structures of the same cardinality and $\mathcal{A} \equiv \mathcal{B}$, then $\mathcal{A} \approx \mathcal{B}$. (cf. Theorem 20.4.)*

Proof. Suppose card $\mathcal{A} = \kappa$. Let $A = \{a_\delta \mid \delta < \kappa\}$ and $B = \{b_\delta \mid \delta < \kappa\}$. A set $\{\langle c_\delta, d_\delta \rangle \mid \delta < \kappa\}$ is defined by transfinite recursion. Fix $\delta < \kappa$ and assume $\{\langle c_\gamma, d_\gamma \rangle \mid \gamma < \delta\}$ has already been defined so that

$$\langle \mathcal{A}, c_\gamma \rangle_{\gamma < \delta} \equiv \langle \mathcal{B}, d_\gamma \rangle_{\gamma < \delta}.$$

(When $\delta = 0$, the above assumption reduces to $\mathcal{A} \equiv \mathcal{B}$.)

Case 1. δ is even. Let c_δ be the member of $A - \{c_\gamma \mid \gamma < \delta\}$ with least subscript. Let

$$p = \{F(x_1) \mid \langle \mathcal{A}, c_\gamma \rangle_{\gamma < \delta} \models F(c_\delta)\}.$$

Then $p \in S_1 T(\langle \mathcal{A}, c_\gamma \rangle_{\gamma < \delta})$. Let q be the result of replacing each occurrence of $\underline{c}_\gamma$ in p by $\underline{d}_\gamma$ for all $\gamma < \delta$. Then $q \in S_1 T(\langle \mathcal{B}, d_\gamma \rangle_{\gamma < \delta})$. Since $\mathcal{B}$ is saturated, q is realized in $\mathcal{B}$ by some b; set $d_\delta = b$. Then

$$\langle \mathcal{A}, c_\gamma, c_\delta \rangle_{\gamma < \delta} \equiv \langle \mathcal{B}, d_\gamma, d_\delta \rangle_{\gamma < \delta}.$$

Case 2. δ is odd. Same as case 1 with $\mathcal{A}$ and $\mathcal{B}$ interchanged.

Let $hc_\delta = d_\delta$ for all $\delta < \kappa$. The back-and-forth feature of the definition of h guarantees that h is a one-to-one correspondence between A and B. h is an isomorphism since

$$\langle \mathcal{A}, a \rangle_{a \in A} \equiv \langle \mathcal{B}, ha \rangle_{a \in A}. \qquad \square$$

Theorem 16.3 is a seductive uniqueness result for saturated structures. It suggests that the problem of classifying structures up to elementary equivalence is reducible to the problem of classifying saturated structures up to isomorphism. Unfortunately saturated structures tend to be rare. Theorem 16.4 is the strongest existence result possible in the absence of assumptions having the flavor of the continuum hypothesis. There are two useful detours around this difficulty. The first consists of accepting partially saturated structures in place of saturated ones, a sensible compromise, since Theorem 16.4 provides an abundance of the former. The second consists of initially assuming the continuum hypothesis and then eliminating it from the final result by means of absoluteness results of K. Gődel, A. Levy and J. Shoenfield.

Theorem 16.4. *Let $\mathcal{A}$ be infinite. For each infinite cardinal κ there exists a κ^+-saturated $\mathcal{B} \succ \mathcal{A}$ such that card $\mathcal{B} \leq (\text{card } \mathcal{A})^\kappa$.*

Proof. Similar to 15.1. First consider the simpler problem of elementarily extending $\mathcal{A}$ to $\mathcal{A}^*$ so that $\mathcal{A}^*$ realizes every $p \in S_1 T(\langle \mathcal{A}, y \rangle_{y \in Y})$ for every $Y \subset A$ of cardinality at most κ; let $\{Y_\delta \mid \delta < (\text{card } A)^\kappa\}$ be an enumeration of all such Y's. An elementary chain $\{\mathcal{A}_\delta \mid \delta < (\text{card } A)^\kappa\}$ is defined by recursion.

(i) $\mathcal{A}_0 = \mathcal{A}$.

(ii) $\mathcal{A}_\lambda = \cup \{\mathcal{A}_\delta \mid \delta < \lambda\}$ if λ is a limit ordinal.

(iii) $\mathcal{A}_{\delta+1} \succ \mathcal{A}_\delta$; $\mathcal{A}_{\delta+1}$ realizes every $p \in S_1 T(\langle \mathcal{A}_\delta, y \rangle_{y \in Y_\delta})$; and card $\mathcal{A}_{\delta+1} \leq$ card $\mathcal{A}_\delta \times 2^\kappa$. The existence of $\mathcal{A}_{\delta+1}$ follows from 15.1(2).

Let $\mathcal{A}^* = \cup \{\mathcal{A}_\delta \mid \delta < (\text{card } \mathcal{A})^\kappa\}$. It is readily checked by induction that card $\mathcal{A}_\delta \leq (\text{card } \mathcal{A})^\kappa$. So card $\mathcal{A}^* \leq (\text{card } \mathcal{A})^\kappa$.

Now $\mathcal{B}$ can be obtained as the limit of an elementary chain $\{\mathcal{B}_\delta \mid \delta < \kappa^+\}$ such that:

$$\mathcal{B}_0 = \mathcal{A},$$
$$\mathcal{B}_\lambda = \cup \{\mathcal{B}_\delta \mid \delta < \lambda\},$$
$$\mathcal{B}_{\delta+1} = \mathcal{B}_\delta^* \text{ and card } \mathcal{B}_\delta^* \leq (\text{card } \mathcal{A})^\kappa.$$

The regularity of κ^+ implies $\mathcal{B}$ is κ^+-saturated: if $Y \subset B$ and card $Y \leq \kappa$, then $Y \subset B_\delta$ for some $\delta < \kappa^+$; so if $p \in S_1 T(\langle \mathcal{B}, y \rangle_{y \in Y})$, then p is realized in $\mathcal{B}_{\delta+1}$, hence in $\mathcal{B} \succ \mathcal{B}_{\delta+1}$. $\qquad \square$

The following consequence of 16.4 is needed for the proof of Chang's two-cardinal theorem (Sec. 23).

Corollary 16.5. *Let κ be a regular uncountable cardinal with the property that $2^\rho \leq \kappa$ for every $\rho < \kappa$. Suppose $\mathcal{A}$ is infinite and is of cardinality at most κ. Then there exists a saturated $\mathcal{B} \succ \mathcal{A}$ of cardinality κ.*

Proof. By 7.3 there is no risk in assuming card $\mathcal{A} = \kappa$. If $\kappa = \rho^+$ for some $\rho < \kappa$, then $\mathcal{B}$ exists by 16.4. Suppose $\kappa > \rho^+$ for every $\rho < \kappa$. Let $\{\rho_\delta \mid \delta < \kappa\}$ be a strictly increasing enumeration of the infinite cardinals less than κ. Then $\mathcal{B}$ is the union of an elementary chain $\{\mathcal{B}_\delta \mid \delta < \kappa\}$ such that:

(i) $\mathcal{B}_0 = \mathcal{A}$;
(ii) $\mathcal{B}_\lambda = \cup \{\mathcal{B}_\delta \mid \delta < \lambda\}$ for every limit ordinal $\lambda < \kappa$;
(iii) $\mathcal{B}_{\delta+1} \succ \mathcal{B}_\delta$, $\mathcal{B}_{\delta+1}$ is ρ_δ^+-saturated, and card $\mathcal{B}_{\delta+1} \leq (\text{card } \mathcal{B}_\delta)^{\rho_\delta}$.

$\mathcal{B}_{\delta+1}$ exists by 16.4. The regularity of κ implies that $\mathcal{B}$ is κ-saturated: if $Y \subset B$ and card $Y < \kappa$, then $Y \subset B_\delta$ for some δ such that $\rho_\delta \geq$ card Y; so if $p \in S_1 T(\langle \mathcal{B}, y \rangle_{y \in Y})$, then p is realized in $\mathcal{B}_{\delta+1}$, hence in $\mathcal{B}$. $\kappa^\rho \leq \kappa$ for every $\rho < \kappa$, since κ is regular and $2^\rho \leq \kappa$ for every $\rho < \kappa$. It follows by induction that card $\mathcal{B}_\delta \leq \kappa$ for every $\delta < \kappa$, so card $\mathcal{B} \leq \kappa$.   □

Theorem 16.6. *Let κ be uncountable. Then $\mathcal{A}$ is κ-saturated iff every diagram of the following sort*

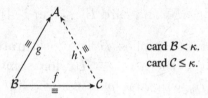

card $\mathcal{B} < \kappa$.
card $\mathcal{C} \leq \kappa$.

can be completed as shown.

Proof. First assume $\mathcal{A}$ has the diagram property. Suppose $Y \subset A$ has cardinality less than κ. The downward Skolem–Löwenheim theorem (11.2) provides a $\mathcal{B} \prec \mathcal{A}$ such that $Y \subset B$ and card $\mathcal{B} < \kappa$. Let

$$p \in S_1 T(\langle \mathcal{A}, y \rangle_{y \in Y}) = S_1 T(\langle \mathcal{B}, y \rangle_{y \in Y}).$$

By 15.1(1) there is a $\mathcal{C} \succ \mathcal{B}$ such that card $\mathcal{C} < \kappa$ and p is realized in $\mathcal{C}$ by some c. Then hc realizes p in $\mathcal{A}$.

Now assume $\mathcal{A}$ is κ-saturated. There is no harm in supposing f and g are inclusion maps. Let $C - B = \{c_\delta \mid \delta < \kappa\}$. The restriction of h to $\mathcal{B}$ is g. hc_δ is defined by recursion. Fix δ and suppose

$$\langle \mathcal{C}, b, c_\gamma \rangle_{b \in B, \gamma < \delta} \equiv \langle \mathcal{A}, b, hc_\gamma \rangle_{b \in B, \gamma < \delta}.$$

Let $p = \{F(x) \mid \langle \mathcal{C}, b, c_\gamma \rangle_{b \in B, \gamma < \delta} \models F(\underline{c_\delta})\}$. Let q be the result of replacing every occurrence of $\underline{c_\gamma}$ in p by $\underline{hc_\gamma}$ for every $\gamma < \delta$. Since $\mathcal{A}$ is κ-saturated, q is realized in $\mathcal{A}$ by some a; set $hc_\delta = a$.   □

Exercise 16.7. $\mathcal{B}$ is said to be finitely generated if there exists a finite $Y \subset B$ such that $\mathcal{B}$ is the least substructure of $\mathcal{B}$ whose universe contains Y. Assume each finite subset of A is contained in the universe of some finitely generated, elementary substructure of $\mathcal{A}$. Show $\mathcal{A}$ is ω-saturated iff every diagram of the following sort

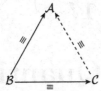

$\mathcal{B}$ is finitely generated.
card $\mathcal{C} \le \omega$.

can be completed as shown.

Exercise 16.8. Let $\mathcal{A} = \langle A, \le \rangle$ be a linear ordering of singular cardinality. Show $\mathcal{A}$ is not saturated.

Exercise 16.9 (H. J. Keisler, S. Kochen). Suppose $\mathcal{A} \equiv \mathcal{B}$. Show that exist $\mathcal{A}^*$ and $\mathcal{B}^*$ such $\mathcal{A} \prec \mathcal{A}^*$, $\mathcal{B} \prec \mathcal{B}^*$ and $\mathcal{A}^* \approx \mathcal{B}^*$.

Section 17

Elimination of Quantifiers for Real Closed Fields

One of the classical applications of model theory to algebra is Tarski's elimination of quantifiers for the theory of real closed fields. Tarki's proof was based on an extension of Sturm's algorithm. The proof below is necessarily less constructive, since it is based on a saturated model criterion for model completeness, but it does give the algebraic details an organization that can be imposed on other theories of fields.

Throughout this section T has no finite models.

Theorem 17.1. *T is model complete iff every diagram of the following sort*

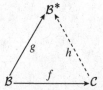

$\mathcal{B}, \mathcal{B}^*, \mathcal{C} \models T.$

$\mathcal{B}^*$ is (card $\mathcal{C}$)$^+$-saturated.

can be completed as shown.

Proof. First assume T is model complete. Then f and g are elementary, so h exists by 16.6.

Now assume T has the above diagram property. It follows from 16.4 that every diagram of the following sort

58

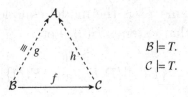

$$\mathcal{B} \models T.$$
$$\mathcal{C} \models T.$$

can be completed as shown. Then T is model complete by 10.4. □

Theorem 17.1 is derived from a criterion given by S. Kochen [Ko1] for model completeness. In Kochen's version, $\mathcal{B}^*$ is replaced by an ultrapower of $\mathcal{B}$. For applications of model theory to algebra, the notion of partially saturated extension seems more germane than the notion of ultrapower. Theorem 17.2 provides the most direct method for establishing completeness, model completeness and elimination of quantifiers for various theories of fields. The directness of 17.2 will be exploited in Sec. 40 to find simple axioms for the theory of differentially closed fields of characteristic 0.

$\mathcal{B}(c)$ denotes a simple extension of $\mathcal{B}$; i.e. $\mathcal{B}(c)$ is the least substructure of $\mathcal{B}(c)$ whose universe contains $B \cup \{c\}$.

Theorem 17.2 (L. Blum). *Let T and T^* be theories in the same language such that $T \subset T^*$, T is universal, and every model of T can be extended to some model of T^*. Then T^* is the model completion of T iff every diagram of the following sort*

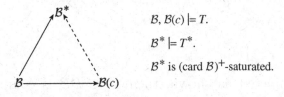

$$\mathcal{B}, \mathcal{B}(c) \models T.$$
$$\mathcal{B}^* \models T^*.$$
$$\mathcal{B}^* \text{ is } (\text{card } \mathcal{B})^+\text{-saturated.}$$

can be completed as shown.

Proof. First assume T^* is the model completion of T. Let $\mathcal{D}$ be a model of T^* that extends $\mathcal{B}(c)$. Let

$$p = \{F(x) | \langle \mathcal{D}, b \rangle_{b \in \mathrm{B}} \models F(\underline{c})\}.$$

$T^* \cup D\mathcal{B}$ is complete, and $p \in S_1(T^* \cup D\mathcal{B})$. The partial saturation of $\mathcal{B}^*$ implies p is realized in $\mathcal{B}^*$ by some b^*; map c to b^*.

Now assume T^* has the given diagram property. Then T^* has a stronger property: every diagram of the following sort

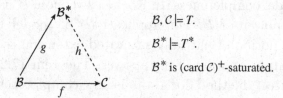

$\mathcal{B}, \mathcal{C} \models T$.

$\mathcal{B}^* \models T^*$.

$\mathcal{B}^*$ is $(\mathrm{card}\, \mathcal{C})^+$-saturated.

can be completed as shown. Let $C - f[B] = \{c_\delta | \delta < \kappa\}$. Define the chain $\{\mathcal{C}_\delta | \delta < \kappa\}$ by: $\mathcal{C}_0 = f[B]$, $\mathcal{C}_{\delta+1} = \mathcal{C}_\delta(c_\delta)$, and $\mathcal{C}_\lambda = \cup \{\mathcal{C}_\delta | \delta < \lambda\}$ when λ is a limit ordinal. Then $h \colon \mathcal{C} \to \mathcal{B}^*$ is defined by κ consecutive uses of the given diagram property.

Let $\mathcal{B}$ be a model of T, and let $\mathcal{C} \supset \mathcal{B}$ and $\mathcal{D} \supset \mathcal{B}$ be models of T^*. By 16.4 there is a $(\mathrm{card}\, \mathcal{D})^+$-saturated $\mathcal{C}^* \succ \mathcal{C}$. If the diagram

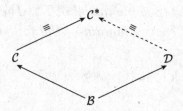

can be completed as shown, then $T^* \cup D\mathcal{B}$ is complete, and consequently, T^* is the model completion of T. $\mathcal{D}$ is a model of T, since $T \subset T^*$. So the stronger diagram property supplies an h such that

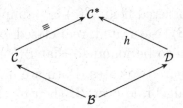

To see that h is elementary, observe that 17.1 and the stronger diagram property imply T^* is model complete. $\qquad\square$

The theory of ordered fields (QF) is the theory of fields (TF) augmented by a 2-place relation symbol $<$ and the following axioms:

(1) $(x) \sim (x < x)$,
(2) $(x)(y)(z)[x < y \,\&\, y < z \to x < z]$,
(3) $(x)(y)[x < y \lor x = y \lor y < x]$,
(4) $(x)(y)[0 < x \,\&\, 0 < y \to 0 < x \cdot y]$,
(5) $(x)(y)(z)[x < y \to x + z < y + z]$.

The theory of real closed fields (RCF) is OF augmented by the axioms:

(6) $(x)(Ey)[0 < x \to x = y \cdot y]$,
(7n) $(x_1) \cdots (x_n)(Ey)[y^n + x_1 y^{n-1} + \cdots + x_n = 0]$

for each odd $n > 0$.

Theorem 17.3 (A. Robinson). *The theory of real closed fields is the model completion of the theory of ordered fields.*

Proof. By 17.2 it suffices to complete the following diagram

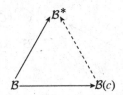

as shown. $\mathcal{B}$ is an ordered field, $\mathcal{B}(c)$ is a simple extension of $\mathcal{B}$, and $\mathcal{B}^*$ is a (card $\mathcal{B}$)$^+$-saturated, real closed extension of $\mathcal{B}$. An element $b^* \in B$ has to be found so that $\mathcal{B}(b^*)$ is isomorphic to $\mathcal{B}(c)$ over $\mathcal{B}$. There are two cases. Both are handled by Sturm's algorithm. If c is algebraic over $\mathcal{B}$, then b^* exists, since every simple algebraic (ordered) extension of an ordered field $\mathcal{B}$ is contained in every real closed extension of $\mathcal{B}$. Suppose c is transcendental over $\mathcal{B}$. Let S be the set of all formulas of the form $f(x) > 0$, where $f(x)$ is a polynomial over $\mathcal{B}$ and $f(c) > 0$.

S is consistent with $T(\langle \mathcal{B}^*, b \rangle_{b \in B})$, since every finite subset of S can be satisfied in every real closed extension of $\mathcal{B}$. Extend S to some $p \in S_1 T(\langle \mathcal{B}^*, b \rangle_{b \in B})$. Then any realization of p in $\mathcal{B}^*$ can serve as b^*. $\qquad \square$

Corollary 17.4 (A. Tarski). *The theory of real closed fields admits elimination of quantifiers.*

Proof. By 13.2 and 17.3. $\qquad \square$

In Sec. 40 it will be shown, by an argument having the same general outline as that of 17.3 and 17.4, that the theory of differentially closed fields of characteristic 0 admits elimination of quantifiers.

Exercise 17.5. T admits elimination of quantifiers iff every diagram of the following sort

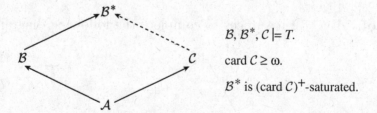

$\mathcal{B}, \mathcal{B}^*, \mathcal{C} \models T$.

card $\mathcal{C} \geq \omega$.

$\mathcal{B}^*$ is (card $\mathcal{C}$)$^+$-saturated.

can be completed as shown.

Exercise 17.6 (L. Blum). *Let T be the model completion of some universal theory. Show there exists a theory $T^* = T$ such that every member of T^* is of the form*

$$(x_1) \ldots (x_n)(Ey)F(x_1, \ldots, x_n, y),$$

where $F(x_1, \ldots, x_n, y)$ has no quantifiers.

Exercise 17.7 (A. Robinson). *A Z-group is an ordered, additive Abelian group with a least positive element (1) that satisfies*

$$(x)(Ey)(Ez)[x = ny + z \,\&\, 0 \leq z < n]$$

for all $n > 0$. Show the theory of Z-groups is model complete.

Exercise 17.8. Let F be a sentence in the language of linear ordering (LO). Suppose F is true in infinitely many finite linear orderings. Show F is true in all but finitely many finite orderings.

Exercise 17.9. Show the theory of finite linear orderings is decidable.

Section 18

Omitting a Type

A structure $\mathcal{A}$ is said to omit an n-type p if no member of A^n realizes p in $\mathcal{A}$. The results of Sec. 17 were straightforward because of the great ease with which structures can be extended to realize n-types. Another class of results, necessarily deeper, employs constructions in which structures are extended and selected n-types are omitted. Clearly it is more difficult to omit than it is to realize, since the act of omitting requires that every element of the extension be worried over. (A not well-known model theorist once remarked: "Any fool can realize a type, but it takes a model theorist to omit one.") In this and future sections a variety of techniques for omitting a type will be developed.

Let T be complete and let $p \in S_n T$. p is principal if some $F(x_1, \ldots, x_n) \in p$ has the property that

$$T \vdash F(x_1, \ldots, x_n) \to G(x_1, \ldots, x_n)$$

for every $G(x_1, \ldots, x_n) \in p$; $F(x_1, \ldots, x_n)$ is said to generate p. (The principal p's of $S_n T$ correspond to the isolated points of $S_n T$ viewed as the Stone space of the Boolean algebra $B_n T$; the generators of the principal p's correspond to the atoms of $B_n T$.) Since T is complete, every model of T realizes every principal p.

Theorem 18.1 (A. Ehrenfeucht). *If T is countable and $p \in$ $S_n T$ is not principal, then T has a model which omits p. (cf. Theorem 24.2.)*

Proof. In the style of the Henkin construction of 7.1. Let $n = 1$ for the sake of notational simplicity. $\{c_i \mid i < w\}$ is a sequence of individual constants, none of which occur in the language of T. $\{G_j(x) \mid j < w\}$ is an enumeration of all formulas (in the language generated by adding the c_i's to the language of T) whose only free variable is x. Let $h : w \to w$ be such that:

(i) $j < i$ implies $hj < hi$;

(ii) $j \leq i$ implies c_{hi} does not occur in $G_j(x)$. The i-th Henkin axiom H_i is

$$(Ex)G_i(x) \to G_i(c_{hi}).$$

An expanding sequence $\{T_i \mid i < w\}$ of theories is defined by induction. $T_0 = T$.

(1) Assume T_{2i} is consistent and c_{hi} does not occur in T_{2i}. Let T_{2i+1} be $T_{2i} \cup \{H_i\}$. Then T_{2i+1} is consistent as in the proof of 7.1.

(2) Assume T_{2i+1} is consistent and

$$T_{2i+1} = T \cup \{K(c_{i1}, \ldots, c_{in}, c_i)\},$$

where $i \neq ij$ when $1 \leq j \leq n$. If there exists an $F(x) \in p$ such that $T_{2i+1} \cup \{\sim F(c_i)\}$ is consistent, then let T_{2i+2} be $T_{2i+1} \cup \{\sim F(c_i)\}$ for some such $F(x)$. Suppose for the sake of a reductio and absurdum that $T_{2i+1} \cup \{\sim F(c_i)\}$ is inconsistent for every $F(x) \in p$. Then

$$T \vdash K(c_{i1}, \ldots, c_{in}, c_i) \to F(c_i)$$

for every $F(x) \in p$. Since $c_{ij}(1 \leq j \leq n)$ and c_i do not occur in T,

$$T \vdash (Ex_1) \cdots (Ex_n)K(x_1, \ldots, x_n, x) \to F(x)$$

for every $F(x) \in p$. If $(Ex_1) \cdots (Ex_n) K(x_1, \ldots, x_n, x) \notin p$, then $\sim (Ex_1) \cdots (Ex_n) K(x_1, \ldots, x_n, x) \in p$,

$$T \vdash \sim (Ex_1) \cdots (Ex_n) K(x_1, \ldots, x_n, x),$$

and T_{2i+1} is inconsistent. If $(Ex_1) \cdots (Ex_n) K(x_1, \ldots, x_n) \in p$, then p is principal.

Let $T_\infty = \cup \{ T_i \mid i < w \}$. The Henkin construction is finished by choosing some maximal consistent extension S of T_∞. As in 7.1, S defines a model of T whose members are equivalence classes of the form $[\underline{c_i}]$. No $[\underline{c_i}]$ realizes p since $\sim F(\underline{c_i}) \in T_{2i+1} \subset S$ for some $F(x) \in p$. $\qquad \square$

The assumption of countability is 18.1 in essential.

T is κ-categorical if all models of T of cardinality κ are isomorphic.

Corollary 18.2 (C. Ryll–Nardjewski). *Let T be a countable, complete theory without any finite models. Then T is ω-categorical iff $S_n T$ is finite for all n iff every countable model of T is saturated.*

Proof. Suppose $S_n T$ is infinite. Then $B_n T$ is infinite. Every infinite Boolean algebra $\mathcal{B}$ has a nonprincipal, maximal dual ideal; one way of obtaining it is to maximize the dual ideal generated by the complements of the atoms of $\mathcal{B}$. So $S_n T$ has a nonprincipal p. By 18.1 there is a countable model of T that omits p. By 15.1 there is a countable model of T that realizes p. So T is not ω-categorical.

If every countable model of T is saturated, then T is ω-categorical by 16.3.

Suppose $S_n T$ is finite for every n. Then every $p \in S_n T$ is principal for every n. Let $\mathcal{A}$ be a countable model of T. Then every n-type realized in $\mathcal{A}$ is principal. Fix

$$p \in S_1 T(\langle \mathcal{A}, a_i \rangle_{1 \leq i \leq n})$$

with the intent of showing p is realized in $\mathcal{A}$. Let

$$p^* = \{G(x_1, \ldots, x_n, x) \mid G(\underline{a}_1, \ldots, \underline{a}_n, x) \in p\}.$$

$p^* \in S_{n+1} T$ and so must be principal; let $H(x_1, \ldots, x_n, x) \in p^*$ generate p^*. Then $H(\underline{a}_1, \ldots, \underline{a}_n, x) \in p$ and generates p. Since p is principal, it is realized in every model of $T(\langle \mathcal{A}, a_i \rangle_{1 \leq i \leq n})$. $\square$

It will be shown in Sec. 37 that if a countable theory T is κ-categorical for some uncountable κ, then every uncountable model of T is saturated. The proof will require a type omitting argument much more intricate than that of 18.1.

Proposition 18.3. *Let $\mathcal{A}$ be saturated, $Y \subset A$, card $Y <$ card A, and $p \in S_n T(\langle \mathcal{A}, y \rangle_{y \in Y})$. Then p is realized in $\mathcal{A}$.*

Proof. By induction on n. Suppose $n > 1$ and $p_n \in S_n T(\langle \mathcal{A}, y \rangle_{y \in Y})$. Let

$$p_{n-1} = \{(Ex_n)F(x_1, \ldots, x_n) \mid F(x_1, \ldots, x_n) \in p_n\}.$$

Then $p_{n-1} \in S_{n-1} T(\langle \mathcal{A}, y \rangle_{y \in Y})$, so p_{n-1} is realized in $\mathcal{A}$ by some $\langle a_1, \ldots, a_{n-1} \rangle$. Let

$$p_1 = \{F(\underline{a}_1, \ldots, \underline{a}_{n-1}, x) \mid F(x_1, \ldots, x_{n-1}, x_n) \in p_n\}.$$

Then $p_1 \in S_1 T(\langle \mathcal{A}, y, a_1, \ldots, a_{n-1} \rangle_{y \in Y})$, so p_1 is realized in $\mathcal{A}$ by some a_n. Clearly $\langle a_1, \ldots, a_n \rangle$ realizes p_n. $\square$

Lemma 18.4. *Let T be a countable, complete theory without any finite models. Then T has a countable saturated model iff $S_n T$ is countable for every n.*

Proof. First suppose T has a countable saturated model $\mathcal{A}$. By 18.3 every $p \in S_n T$ is realized in $\mathcal{A}$. Then $S_n T$ is countable, since A^n is countable.

Now suppose S_nT is countable for every n. It follows that $S_nT(\langle \mathcal{B}, y \rangle_{y \in Y})$ is countable for every n, every model $\mathcal{B}$ of T, and every finite $Y \subset B$: Let $Y = \{y_1, \ldots, y_m\}$. If $S_nT(\langle \mathcal{B}, y \rangle_{y \in Y})$ is uncountable, then $S_{m+n}T$ is uncountable, since there is a one–one map of the former into the latter induced by the passage from

$$F(\underline{y}_1, \ldots, \underline{y}_m, x_1, \ldots, x_n) \quad \text{to} \quad F(x_1, \ldots, x_m, \ldots, x_{m+n}).$$

Let $\mathcal{B}$ be a countable model of T, and let $\{Y_i \mid i < w\}$ be an enumeration of all finite subsets of B. An elementary chain $\{\mathcal{B}_i \mid i < w\}$ is defined so that:

(i) $\mathcal{B}_0 = \mathcal{B}$.

(ii) $\mathcal{B}_{i+1} \succ \mathcal{B}_i$. $\mathcal{B}_{i+1}$ realizes every $p \in S_1T(\langle \mathcal{B}, y \rangle_{y \in Y_i})$. $\mathcal{B}_{i+1}$ is countable. The existence of $\mathcal{B}_{i+1}$ follows from 15.1 and the fact that $S_1T(\langle \mathcal{B}, y \rangle_{y \in Y})$ is countable.

Let $\mathcal{C} = \cup \{\mathcal{B}_i \mid i < w\}$. Then $\mathcal{C} \succ \mathcal{B}$ by 10.3; every $p \in S_1T(\langle \mathcal{B}, y \rangle_{y \in Y})$ is realized in $\mathcal{C}$ for every finite $Y \in \mathcal{B}$; and $\mathcal{C}$ is countable.

The desired countable saturated model $\mathcal{A}$ is the limit of an elementary chain $\{\mathcal{A}_i \mid i < w\}$ defined so that:

(iii) $\mathcal{A}_0$ is a countable model of T.

(iv) $\mathcal{A}_{i+1} \succ \mathcal{A}_i$. Every $p \in S_1T(\langle \mathcal{A}_i, y \rangle_{y \in Y})$ is realized in $\mathcal{A}_{i+1}$ for every finite $Y \subset A_i$. $\mathcal{A}_{i+1}$ is countable. $\square$

Section 19

ω-Stable Theories

Let $S_1\mathcal{A}$ denote $S_1T(\langle\mathcal{A}, a\rangle_{a\in A})$. Let T be a countable theory without any finite models. T is κ-stable (M. Morley) if card $S_1\mathcal{A} = \kappa$ whenever $\mathcal{A}$ is a model of T of cardinality κ. The notion of ω-stability is absolute; in fact it is Π_1^1.

Lemma 19.1 (M. Morley). *If T is ω-stable, then T is κ-stable for every $\kappa \geq \omega$.*

Proof. Suppose $\mathcal{A} \vdash T$ and card $\mathcal{A} <$ card $S_1\mathcal{A}$. Let $p, q, \ldots \in S_1\mathcal{A}$, and let $F(x), G(x), \ldots$ be formulas in the language of $T(\langle\mathcal{A}, a\rangle_{a\in A})$. $F(x)$ is *large* if

$$\text{card}\{p \mid F(x) \in p\} > \text{card } \mathcal{A}.$$

(In other words, $F(x)$ defines a large neighborhood of the Stone space $S_1\mathcal{A}$.) Clearly $x = x$ is large. Suppose $F(x)$ is large with the intent of finding a $G(x)$ such that both $F(x)$ & $G(x)$ and $F(x)$ & $\sim G(x)$ are large. Let Q be the set of all p such that for some $H(x)$,

$$F(x) \, \& \, H(x) \in p \quad \text{and} \quad F(x) \, \& \, H(x) \text{ is not large}.$$

Then card $Q \leq (\text{card } \mathcal{A})^2 < \text{card } \{p \mid F(x) \in p\}$. Choose

$$q_1, q_2 \in \{p \mid F(x) \in p\} - Q$$

69

such that $q_1 \neq q_2$. Choose $G(x) \in q_1$ so that $\sim G(x) \in q_2$. Then both $F(x) \& G(x)$ and $F(x) \& \sim G(x)$ are large.

A sequence $\{F_j^i \mid j < 2^i < \omega\}$ of large formulas is defined by induction. F_0^0 is $x = x$. Assume F_j^i has been defined. Choose $G(x)$ so that both $F_j^i \& G(x)$ and $F_j^i \& \sim G(x)$ are large. Let F_{2j}^{i+1} be $F_j^i \& G(x)$, and F_{2j+1}^{i+1} be $F_j^i \& \sim G(x)$. It follows from 11.2 that there exists a countable $\mathcal{B} \prec \mathcal{A}$ such that B contains all the members of A mentioned in the F_j^i's. Let $t : \omega \to \omega$ have the property that $t(0) = 0$ and $t(i+1) \in \{2t(i), 2t(i)+1\}$. Then $\{F_{t(i)}^i \mid i < \omega\}$ is consistent with $T(\langle \mathcal{B}, b \rangle_{b \in B})$, and so can be extended to some $p_t \in S_1 \mathcal{B}$. If $t \neq t'$, then $p_t \neq p_{t'}$ since there is a $G(x)$ such that $G(x) \in p_t$ and $\sim G(x) \in p_{t'}$. But then $S_1 \mathcal{B}$ is uncountable. $\qquad\square$

The argument of 19.1 leads to a more general result: For any Boolean algebra $\mathcal{B}$, let $S\mathcal{B}$ be the Stone space of $\mathcal{B}$, i.e. the set of all maximal dual ideals of $\mathcal{B}$; if SC is countable for every countable $\mathcal{C} \subset \mathcal{B}$, then card $S\mathcal{B}$ = card $\mathcal{B}$.

S. Shelah [Shl] has extended 19.1 as follows: suppose T is not ω-stable but is κ-stable for some $\kappa > \omega$; then the set of all κ such that T is κ-stable is either $\{\kappa \mid \kappa^\omega = \kappa\}$ or $\{\kappa \mid \kappa \geq 2^\omega\}$.

Lemma 19.1 states the single most important fact about ω-stable theories. In this section, it will be used to show ω-stable theories have an abundance of saturated models. In Sec. 35 it will be used to show that uncountable models of ω-stable theories are rich in indiscernibles. A simple, but somewhat misleading, example of an ω-stable theory is the theory of algebraically closed fields of characteristic 0 (ACF$_0$), misleading because it was shown in Sec. 16 that every uncountable model of ACF$_0$ is saturated yet there are ω-stable theories with unsaturated models in every uncountable cardinality. The precise definition of indiscernible will come later, but even now it is not hard to grasp what is meant by saying that the members of a transcendence base for a model $\mathcal{A}$ of ACF$_0$ are indiscernible in $\mathcal{A}$.

The theory of linear ordering (LO) is not ω-stable, since there exist uncountably many Dedekind cuts in the ordering of the rationals. LO is an excellent example of a non-ω-stable theory; in fact, it will be shown in Sec. 35 that every theory possessing a model with a definable, infinite linear ordering fails to be ω-stable for the same reason LO does. At this writing the least misleading example of an ω-stable theory is the theory of differentially closed fields of characteristic 0 (DCF$_0$). The ω-stability of DCF$_0$ will be exploited in Sec. 41 to show that every differential field of characteristic 0 has a unique, prime differential closure.

Theorem 19.2. *Suppose $\mathcal{A}$ is infinite and $T\mathcal{A}$ is ω-stable. If $\rho \leq \operatorname{card}\mathcal{A}$ and is regular, then there exists a ρ-saturated $\mathcal{B} \succ \mathcal{A}$ of the same cardinality as $\mathcal{A}$.*

Proof. An elementary chain $\{\mathcal{B}_\delta \,|\, \delta \leq \rho\}$ is defined so that:

$\mathcal{B}_0 = \mathcal{A}$;

$\mathcal{B}_{\delta+1}$ realizes every $p \in S_1\mathcal{B}_\delta$;

$\mathcal{B}_\lambda = \cup\{\mathcal{B}_\delta \,|\, \delta < \lambda\}$ when λ is a limit ordinal.

Let $\mathcal{B} = \mathcal{B}_\rho$. By 15.1(2) and 19.1, $\mathcal{B}_{\delta+1}$ can be chosen so that card $\mathcal{B}_{\delta+1} = $ card $\mathcal{B}_\delta$; so card $\mathcal{B} = $ card $\mathcal{A}$. The regularity of ρ insures that $\mathcal{B}$ is ρ-saturated. $\square$

Corollary 19.3 (M. Morley). *If T is ω-stable and κ is regular, then T has a saturated model of cardinality κ.*

It will be shown at the end of Sec. 35 that ω-stable theories have saturated models in *all* infinite cardinalities.

Exercise 19.4. $\mathcal{A}$ is a special model of T if there exists an elementary chain $\{\mathcal{A}_\gamma \,|\, \gamma < \alpha\}$ of saturated models such that $\mathcal{A} = \cup\{\mathcal{A}_\gamma \,|\, \gamma < \alpha\}$. Suppose T is ω-stable and $\kappa \geq \omega$. Show T has a special model of cardinality κ. (cf. Theorem 35.10.)

Section 20

Homogeneous Structures

Let $\mathcal{A}$ be infinite, X, $Y \subset \mathcal{A}$, and $f\colon X \to Y$ be onto. f is an elementary partial automorphism of $\mathcal{A}$ if

$$\langle \mathcal{A}, x \rangle_{x \in X} \equiv \langle \mathcal{A}, fx \rangle_{x \in X}.$$

Card f is defined to be card X. f is immediately extendible if for each $a \in A$ there is a $b \in A$ such that

$$\langle \mathcal{A}, x, a \rangle_{x \in X} \equiv \langle \mathcal{A}, fx, b \rangle_{x \in X}.$$

$\mathcal{A}$ is homogeneous if every elementary partial automorphism of $\mathcal{A}$, of lesser cardinality than $\mathcal{A}$, is immediately extendible. Countable homogeneous structures will be needed in Sec. 22 for the proof of Vaught's two-cardinal theorem.

Lemma 20.1. *$\mathcal{A}$ is homogeneous iff every elementary partial automorphism of $\mathcal{A}$ of cardinality less than $\mathcal{A}$ can be extended to an automorphism of $\mathcal{A}$.*

Proof. Let $f\colon X \to Y$ be a partial automorphism of $\mathcal{A}$ of cardinality less than $\mathcal{A}$. Let $A - X = \{x_\delta \mid \delta < \text{card } \mathcal{A}\}$ and $A - Y = \{y_\delta \mid \delta < \text{card } \mathcal{A}\}$. A set $\{\langle c_\delta, d_\delta \rangle \mid \delta < \text{card } \mathcal{A}\}$ is defined by recursion. Fix δ and assume $\{\langle c_\gamma, d_\gamma \rangle \mid \gamma < \delta\}$ has already been defined so that

$$\langle \mathcal{A}, x, c_\gamma \rangle_{x \in X, \gamma < \delta} \equiv \langle \mathcal{A}, fx, d_\gamma \rangle_{x \in X, \gamma < \delta}.$$

Case 1. δ is even. Let c_δ be the member of $(A - X) - \{c_\gamma \mid \gamma < \delta\}$ with least subscript. Since $\mathcal{A}$ is homogeneous, there is a $y \in (A - Y) - \{d_\gamma \mid \gamma < \delta\}$ such that

$$\langle \mathcal{A}, x, c_\gamma, c_\delta \rangle_{x \in X, \gamma < \delta} \equiv \langle \mathcal{A}, fx, d_\gamma, y \rangle_{x \in X, \gamma < \delta}.$$

Set $d_\delta = y$.

Case 2. δ is odd. Same as case 1 with X and Y interchanged: d_δ is the member of $(A - Y) - \{d_\gamma \mid \gamma < \delta\}$ with least subscript etc.

Extend $f: X \to Y$ by setting $fc_\delta = d_\delta$ for all $\delta <$ card $\mathcal{A}$. The back-and-forth feature of the above construction guarantees that f maps $\mathcal{A}$ onto $\mathcal{A}$. $\mathcal{A}$ is an automorphism since

$$\langle \mathcal{A}, a \rangle_{a \in A} \equiv \langle \mathcal{A}, fa \rangle_{a \in A}.$$

$\square$

Let $\mathcal{A}$ be infinite. $\mathcal{A}$ is universal if there exists an $f: \mathcal{B} \overset{\equiv}{\to} \mathcal{A}$ whenever $\mathcal{B} \equiv \mathcal{A}$ and card $\mathcal{B} \leq$ card $\mathcal{A}$.

Theorem 20.2 (M. Morley, R. Vaught). $\mathcal{A}$ *is saturated iff* $\mathcal{A}$ *is homogeneous and universal.*

Proof. Suppose $\mathcal{A}$ is saturated. To see that $\mathcal{A}$ is homogeneous let $f: X \to Y$ be such that

$$\langle \mathcal{A}, x \rangle_{x \in X} \equiv \langle \mathcal{A}, fx \rangle_{x \in X}$$

and card $X <$ card $\mathcal{A}$. Suppose $a \in A$ and $p \in S_1(\langle \mathcal{A}, x \rangle_{x \in X})$ is the 1-type realized by a in $\langle \mathcal{A}, x \rangle_{x \in X}$. Then $p \in S_1 T(\langle \mathcal{A}, fx \rangle_{x \in X})$; so p is realized in $\langle \mathcal{A}, fx \rangle_{x \in X}$ by some b. Clearly

$$\langle \mathcal{A}, x, a \rangle_{x \in X} \equiv \langle \mathcal{A}, fx, b \rangle_{x \in X}.$$

$\mathcal{A}$ is universal by the argument used to complete the diagram in 16.6 and 16.7.

Now suppose $\mathcal{A}$ is universal and homogeneous.

To see that $\mathcal{A}$ is saturated let $P \in S_1 T(\langle \mathcal{A}, y \rangle_{y \in Y})$ for some $Y \subset A$ such that card $Y <$ card A. By 15.1(1), there is a $\mathcal{C} \succ \mathcal{A}$

with the property that card $C = $ card A and p is realized in $\langle C, y \rangle_{y \in Y}$ by some c. Since A is universal, there is an $f: C \overset{\equiv}{\to} A$. So

$$\langle A, fy \rangle_{y \in Y} \equiv \langle A, y \rangle_{y \in Y},$$

and fc realizes p in $\langle A, fy \rangle_{y \in Y}$. Since A is homogeneous, there is $b \in A$ such that

$$\langle A, fy, fc \rangle_{y \in Y} \equiv \langle A, y, b \rangle_{y \in Y}.$$

Then b realizes p in $\langle A, y \rangle_{y \in Y}$. $\qquad\square$

Lemma 20.3. *Suppose $A \equiv B$, card $A \leq$ card B, and B is homogeneous. Suppose for every n and every $p \in S_n TA$: if p is realized in A, then p is realized in B. Then there exists an $f: A \overset{\equiv}{\to} B$.*

Proof. Let $X \subset A$. An induction on card X will show

$$\langle A, x \rangle_{x \in X} \equiv \langle B, fx \rangle_{x \in X}$$

for some $f: X \to B$. If card X is finite, then X realizes some n-type of TA which is also realized in B. Assume card X is infinite. Let $X = \{ x_\delta \mid \delta < $ card $X \}$. $f: X \to B$ is defined by recursion on δ. Fix $\delta < $ card X and suppose $f: \{ x_\gamma \mid \gamma < \delta \} \to B$ has already been defined so that

$$\langle A, x_\gamma \rangle_{\gamma < \delta} \equiv \langle B, fx_\gamma \rangle_{\gamma < \delta}.$$

Since card $\{ x_\gamma \mid \gamma \leq \delta \} < $ card X, there is by induction a g such that

$$\langle A, x_\gamma \rangle_{\gamma \leq \delta} \equiv \langle B, gx_\gamma \rangle_{\gamma \leq \delta}.$$

Since B is homogeneous, there is a $b \in B$ with the property that

$$\langle B, gx_\gamma, gx_\delta \rangle_{\gamma < \delta} \equiv \langle B, fx_\gamma, b \rangle_{\gamma < \delta}.$$

Set $fx_\delta = b$. $\qquad\square$

Let $\mathcal{A} \equiv \mathcal{B}$. $\mathcal{A}$ and $\mathcal{B}$ are said to realize the same n-types if for all $p \in S_n T\mathcal{A}$: p is realized in $\mathcal{A}$ iff p is realized $\mathcal{B}$.

Theorem 20.4 (H.J. Keisler, M. Morley). *If $\mathcal{A}$ and $\mathcal{B}$ are of the same cardinality, elementarily equivalent, homogeneous and realize the same n-types for all n, then $\mathcal{A} \approx \mathcal{B}$.*

Proof. Let $A = \{a_\delta \mid \delta < \kappa\}$ and $B = \{a_\delta \mid \delta < \kappa\}$. A set $\{\langle c_\delta, d_\delta\rangle \mid \delta < \kappa\}$ is defined by recursion. Fix δ and assume

$$\langle \mathcal{A}, c_\gamma\rangle_{\gamma < \delta} \equiv \langle \mathcal{B}, d_\gamma\rangle_{\gamma < \delta}.$$

Case 1. δ is even. Let c_δ be the member of $A - \{c_\gamma \mid \gamma < \delta\}$ with least subscript. By 20.3 there is an f such that

$$\langle \mathcal{A}, c_\gamma\rangle_{\gamma \leq \delta} \equiv \langle \mathcal{B}, fc_\gamma\rangle_{\gamma \leq \delta}.$$

Since $\mathcal{B}$ is homogeneous, there is a $d \in B$ such that

$$\langle \mathcal{B}, d_\gamma, d\rangle_{\gamma < \delta} \equiv \langle \mathcal{B}, fc_\gamma, fc_\delta\rangle_{\gamma < \delta}.$$

Set $d_\delta = d$.

Case 2. δ is odd. Same as case 1 with $\mathcal{A}$ and $\mathcal{B}$ interchanged. $\square$

Theorem 20.5 (R. Vaught). *Let $\mathcal{A}$ be countably infinite. Then there exists a countable homogeneous $\mathcal{B} \succ \mathcal{A}$.*

Proof. Let C be infinite, and let $f\colon X \to Y$ be a finite, elementary partial automorphism of C.

Suppose $C \prec \mathcal{D}$; f is said to be immediately extendible in $\mathcal{D}$ if for each $c \in C$, there is a $d \in D$ such that

$$\langle C, x, c\rangle_{x \in X} \equiv \langle \mathcal{D}, fx, d\rangle_{x \in X}.$$

A slight modification of the proof of 15.1(2) shows: if C is countable, then there exists a countable $\mathcal{D} \succ C$ such that every finite, elementary partial automorphism of C is immediately extendible in $\mathcal{D}$. Define an elementary chain $\{\mathcal{A}_n \,|\, n < \omega\}$ so that:

$\mathcal{A}_0 = \mathcal{A}$;

$\mathcal{A}_{n+1}$ is countable;

every finite, elementary partial automorphism of $\mathcal{A}_n$ is immediately extendible in $\mathcal{A}_{n+1}$.

Then $\mathcal{B} = \cup\{\mathcal{A}_n \,|\, n < \omega\}$ is a countable, homogeneous elementary extension of $\mathcal{A}$. $\square$

The next proposition will be used to omit a type in Sec. 22.

Proposition 20.6. *Let α be countable, and let $\{\mathcal{A}_\delta \,|\, \delta < \alpha\}$ be an elementary chain of countable, isomorphic, homogeneous structures. Then $\cup\{\mathcal{A}_\delta \,|\, \delta < \alpha\} \approx \mathcal{A}_0$.*

Proof. Let $\mathcal{A}_\alpha = \cup\{\mathcal{A}_\delta \,|\, \delta < \alpha\}$. Then $\mathcal{A}_\alpha \succ \mathcal{A}_0$, $\mathcal{A}_\alpha$ is homogeneous, and $\mathcal{A}_\alpha$ and $\mathcal{A}_0$ realize the same n-types. By 20.4 $\mathcal{A}_\alpha \approx \mathcal{A}_0$. $\square$

Exercise 20.1 (A. Ehrenfeuct). An onto map $f \colon X \to Y$ is called a partial isomorphism between $\mathcal{A}$ and $\mathcal{B}$ if $X \subset A$, $Y \subset B$ and $\langle \mathcal{A}, x \rangle_{x \in X}$ and $\langle \mathcal{B}, fx \rangle_{x \in X}$ satisfy the same quantifierless sentences.

f is immediately extendible if for each $a \in A$ (respectively $b \in B$) there is a $b \in B$ (respectively $a \in A$) such that $\langle \mathcal{A}, x, a \rangle_{x \in X}$ and $\langle \mathcal{A}, fx, b \rangle_{x \in X}$ satisfy the same quantifierless sentences.

Suppose every finite, partial isomorphism between $\mathcal{A}$ and $\mathcal{B}$ is immediately extendible. Show $\mathcal{A} \equiv \mathcal{B}$.

Section 21

The Number
of Countable Models

Throughout this section T is a complete, countable theory without finite models. Let $n(T)$ be the number of isomorphism classes of countable models of T. For each cardinal

$$\kappa \in \{n \mid n \neq 2 \ \& \ 1 \leq n < \omega\} \cup \{w, 2^\omega\},$$

there is a T such that $n(T) = \kappa$. R. Vaught [Va1] has conjectured: $n(T) < 2^\omega$ implies $n(T) \leq \omega$. M. Morley [Mo1] has proved: $n(T) < 2^\omega$ implies $n(T) \leq \omega_1$; his proof employs infinitary logic. A. H. Lachlan [La1] has shown: if T is ω-stable and $n(T) > 1$, then $n(T) \geq \omega$. It will be seen in Sec. 39 that if T is ω_1-categorical and $n(T) > 1$, then $n(T) = \omega$. The rather remarkable result of Vaught (21.5) that $n(T)$ can never be 2 is a corollary of the next theorem.

$\mathcal{A}$ is a weakly saturated model of T if every $p \in S_n T$ is realized in $\mathcal{A}$ for every n.

Theorem 21.1 (J. Rosenstein). *If $1 < n(T) < \omega$ then T has a countable, weakly saturated, unsaturated model. (cf. Exercise 39.13.)*

Proof. $S_n T$ is countable for every n, since $n(T) \leq \omega$. By 18.4 T has a countable saturated model $\mathcal{A}$. Let $\mathcal{A}_1, \ldots, \mathcal{A}_\kappa$ be the countable unsaturated models of T. By 16.3 $n(T) = \kappa + 1$.

Suppose for the sake of a reductio ad absurdum that no $\mathcal{A}_i$ $(1 \leq i \leq \kappa)$ is weakly saturated. Choose ni and $p_i \in S_{ni}T$ so that p_i is not realized in $\mathcal{A}_i$.

For any formula F, let F^j be the result of adding j to the subscript of every free variable of F. Define

$$q_i = \{F^{n1+n2+\cdots+n(i-1)} \mid F \in p_i\}.$$

Then p_i and q_i have the same realizations, but q_i and q_j have no free variables in common when $i \neq j$. Let $q = \cup\{q_i \mid 1 \leq u \leq \kappa\}$. Every q_i is realized in $\mathcal{A}$ by 18.3, so q is realized in $\mathcal{A}$.

Let

$$T^* = T \cup \{F(\underline{c}_1, \ldots, \underline{c}_n) \mid F(x_1, \ldots, x_n) \in q\},$$

where $c_1, \ldots, c_n$ do not occur in T and $n = \Sigma\{ni \mid 1 \leq i \leq \kappa\}$. Let $\mathcal{B}$ be a countable model of T^*. Then $\mathcal{B}$ is a model of T. $\mathcal{B} \not\approx \mathcal{A}_i$ $(1 \leq i \leq \kappa)$ since $\mathcal{B}$ realizes q_i. So $\mathcal{B} \approx \mathcal{A}$. Thus each countable model of T^* consists of $\mathcal{A}$ and a realization of q in $\mathcal{A}$. By 20.2 $\mathcal{A}$ is homogeneous; consequently T^* is ω-categorical. By 18.2 $S_n T^*$ is finite for all n. Hence $S_n T$ is finite for all n, since every member of $S_n T$ extends to some member of $S_n T^*$, and since distinct members of $S_n T$ have distinct extensions. But then $n(T) = 1$ by 18.2.   □

$\mathcal{A}$ is a prime model of T if for every model $\mathcal{B}$ of T, there exists an $f: \mathcal{A} \overset{\equiv}{\to} \mathcal{B}$.

$\mathcal{A}$ is an atomic model of T if for all n, every $p \in S_n T$ realized in $\mathcal{A}$ is principal.

Lemma 21.2 (R. Vaught). *$\mathcal{A}$ is a prime model of T iff $\mathcal{A}$ is a countable atomic model of T.*

Proof. Suppose $\mathcal{A}$ is prime. $\mathcal{A}$ must be countable, since T has some countable model. Suppose $p \in S_n T$ is not principal. By 18.1 there is a model $\mathcal{B}$ of T such that p is not realized in $\mathcal{B}$. Since there is an $f: \mathcal{A} \overset{\equiv}{\to} \mathcal{B}$, p is not realized in $\mathcal{A}$.

Assume $\mathcal{A}$ is countable and atomic, $A = \{a_i \mid i < \omega\}$ and $\mathcal{B}$ is a model of T. An $f: \mathcal{A} \overset{\equiv}{\to} \mathcal{B}$ is defined by induction. Fix $n < \omega$ and assume fa_i has been defined for all $i < n$ so that

$$\langle \mathcal{A}, a_i \rangle_{i<n} \equiv \langle \mathcal{B}, fa_i \rangle_{i<n}.$$

Let $p \in S_n T$ be the n-type realized by $\langle a_0, \ldots, a_n \rangle$ in $\mathcal{A}$. Since $\mathcal{A}$ is atomic, p has a generator $F(x_0, \ldots, x_n)$. Then

$$\langle \mathcal{A}, a_i \rangle_{i<n} \models (Ex)F(\underline{a}_0, \ldots, \underline{a}_{n-1}, x),$$

and so

$$\langle \mathcal{B}, fa_i \rangle_{i<n} \models F(\underline{fa}_0, \ldots, \underline{fa}_{n-1}, \underline{b})$$

for some $b \in B$. Set $fa_n = b$. □

Theorem 21.3 (R. Vaught). *If $\mathcal{A}$ and $\mathcal{B}$ are prime models of T, then $\mathcal{A} \approx \mathcal{B}$.*

Proof. By 21.2 $\mathcal{A}$ and $\mathcal{B}$ are atomic and countable. Let $A = \{a_i \mid i < \omega\}$. A set $\{\langle c_i, d_i \rangle \mid i < \omega\}$ is defined by induction. Fix $n < \omega$ and assume $\{\langle c_i, d_i \rangle \mid i < n\}$ has been defined so that

$$\langle \mathcal{A}, c_i \rangle_{i<n} \equiv \langle \mathcal{B}, d_i \rangle_{i<n}.$$

Case 1. n is even. Let c_n be the member of $A - \{c_i \mid i < n\}$ with least subscript. Let $p \in S_n T$ be the n-type realized by $\langle c_0, \ldots, c_n \rangle$ in $\mathcal{A}$. Since $\mathcal{A}$ is atomic, p has a generator $F(x_0, \ldots, x_n)$. Then

$$\langle \mathcal{A}, c_i \rangle_{i<n} \models (Ex)F(\underline{c}_0, \ldots, \underline{c}_{n-1}, x),$$

and so

$$\langle \mathcal{B}, d_i \rangle_{i<n} \models F(\underline{d}_0, \ldots, \underline{d}_{n-1}, \underline{b})$$

for some $b \in B$. Set $d_n = b$.

Case 2. n is odd. Same as case 1 with $\mathcal{A}$ and $\mathcal{B}$ interchanged.

Define f by $fc_n = d_n$. Then $f: \mathcal{A} \overset{\approx}{\to} \mathcal{B}$. □

Lemma 21.4 (R. Vaught). *If $\mathcal{A}$ is a prime model of T, then $\mathcal{A}$ is homogeneous.*

Proof. By 21.2 $\mathcal{A}$ is countable and atomic. Suppose

$$\langle \mathcal{A}, a_1, \ldots, a_{n-1} \rangle \equiv \langle \mathcal{A}, b_1, \ldots, b_{n-1} \rangle.$$

Let $a \in A$, and let $p \in S_n T$ be the n-type realized by $\langle a_1, \ldots, a_{n-1}, a \rangle$ in $\mathcal{A}$. Since $\mathcal{A}$ is atomic, p has a generator $F(x_1, \ldots, x_n)$. As in 21.3,

$$\langle \mathcal{A}, b_1, \ldots, b_{n-1} \rangle \models F(\underline{b}_1, \ldots, \underline{b}_{n-1}, \underline{b})$$

for some $b \in A$. Then

$$\langle \mathcal{A}, a_1, \ldots, a_{n-1}, a \rangle \equiv \langle \mathcal{A}, b_1, \ldots, b_{n-1}, b \rangle. \qquad \square$$

Corollary 21.5 (R. Vaught). $n(T) \neq 2$.

Proof. Suppose $n(T) = 2$. Let $\mathcal{A}$ and $\mathcal{B}$ be the countable models of T. Since $n(T) \leq \omega$. T has a countable saturated model, let it be $\mathcal{B}$, by 18.4. There exists an $f : \mathcal{A} \overset{\equiv}{\to} \mathcal{B}$ by 20.2. It follows from the downward Skolem–Löwenheim theorem (11.2) that $\mathcal{A}$ is prime. Since $n(T) \neq 1$, there is an n such that $S_n T$ is infinite by 18.2. As in 18.4, some $p \in S_n T$ is not principal. By 21.2 p is not realized in $\mathcal{A}$, so $\mathcal{A}$ is not weakly saturated. But 21.1 requires $\mathcal{A}$ to be weakly saturated. $\qquad \square$

$\mathcal{A}$ is a minimal model of T if there is no $\mathcal{B}$ such that $\mathcal{B} \prec \mathcal{A}$ and $\mathcal{B} \neq \mathcal{A}$.

Theorem 21.6 (R. Vaught). *Suppose T is ω_1-categorical but not ω-categorical. If $\mathcal{A}$ is a prime model of T, then $\mathcal{A}$ is a minimal model of T.*

Proof. Suppose $\mathcal{A}$ is prime but not minimal. Then there exists a $\mathcal{B}$ such that $\mathcal{B} \prec \mathcal{A}$, $\mathcal{B} \neq \mathcal{A}$ and $\mathcal{B} \approx \mathcal{A}$ by 21.3. It follows that if $\mathcal{C}$ is a prime model of T, then there exists a $\mathcal{D}$ such that

$\mathcal{C} \prec \mathcal{D}$, $\mathcal{D} \neq \mathcal{C}$ and $\mathcal{D} \approx \mathcal{C}$. An elementary chain $\{\mathcal{A}_\delta \mid \delta < \omega_1\}$ of countable models of T is defined:

(1) $\mathcal{A}_0$ is prime.
(2) Assume $\mathcal{A}_\delta$ is prime; choose $\mathcal{A}_{\delta+1}$ so that $\mathcal{A}_\delta \prec \mathcal{A}_{\delta+1}$, $\mathcal{A}_{\delta+1} \neq \mathcal{A}_\delta$ and $\mathcal{A}_{\delta+1} \approx \mathcal{A}_\delta$.
(3) Let $\mathcal{A}_\lambda = \cup\{\mathcal{A}_\delta \mid \delta < \lambda\}$ when λ is a limit ordinal. If $\mathcal{A}_\delta$ is prime, hence atomic, for all $\delta < \lambda$, then $\mathcal{A}_\lambda$ is atomic, hence prime by 21.2.

Let $\mathcal{C} = \cup\{\mathcal{A}_\delta \mid \delta < \omega_1\}$. Then $\mathcal{C}$ is an atomic model of T of cardinality ω_1. Since T is not ω-categorical, there is an n and a $p \in S_n T$ such that p is not principal by 18.2. Let $\mathcal{D}$ be a model of T that realizes p and has cardinality ω_1. Then T is not ω_1-categorical. □

It will be seen in Sec. 38 that every ω_1-categorical T has a prime model.

Exercise 21.7. The principal points of $S_n T$ are said to be *dense* in $S_n T$ if for every $F(x_1, \ldots, x_n)$ consistent with T there is a principal $p \in S_n T$ such that $F(x_1, \ldots, x_n) \in p$. Show T has a prime model iff the principal points of $S_n T$ are dense in $S_n T$ for every n.

Exercise 21.8. Suppose T is ω-stable. Show T has a prime model.

Exercise 21.9. Suppose $n(T) > 1$ and every countable model of T is homogeneous. Show $n(T) \geq \omega$.

Exercise 21.10. Suppose $n(T) > \omega$ and every countable model of T is homogeneous. Show $n(T) = 2^\omega$.

Section 22

Vaught's Two-Cardinal Theorem

Let $\mathscr{L}$ be a countable language, and let $R(x)$ be a distinguished formula of $\mathscr{L}$. If $\mathcal{A}$ has the same similarity type as $\mathscr{L}$, then the two-cardinal type of $\mathcal{A}$ is $\langle \kappa, \rho \rangle$, where $\kappa = \text{card } \mathcal{A}$, $\rho = \text{card } R^{\mathcal{A}}$, and $R^{\mathcal{A}} = \{a \mid \mathcal{A} \models R(\underline{a})\}$. (It is customary to call $R^{\mathcal{A}}$ the distinguished subset of $\mathcal{A}$.) The expression

$$\langle \kappa, \rho \rangle \rightarrow \langle \kappa', \rho' \rangle$$

means: for each $\mathcal{A}$ of type $\langle \kappa, \rho \rangle$, there is a $\mathcal{B} \equiv \mathcal{A}$ of type $\langle \kappa', p' \rangle$. Vaught initiated the study of two cardinal theorems by proving $\langle \kappa, \rho \rangle \rightarrow \langle \omega_1, \omega \rangle$ (Theorem 22.6) for all $k > \rho \geq \omega$. Theorem 22.6 will be exploited in Secs. 38 and 39 to develop a notion of dimension for the models of an ω_1-categorical theory. As a rule two-cardinal theorems are proved by omitting a type; sometimes they are used, as in Corollary 24.5, to omit a type. The type omitted in the proof of a two-cardinal theorem is: "x is a new member of the distinguished subset." Thus in the proof of 22.4, $\mathcal{C}_\delta$ is enlarged to $\mathcal{C}_{\delta+1}$ without enlarging the distinguished subset.

Proposition 22.1. *Suppose $\mathcal{A} \prec \mathcal{B}$, card $\mathcal{B} = \omega$, $R^{\mathcal{A}} = R^{\mathcal{B}}$, $b \in B - A$ and $p \in S_n T(\langle B, b_i \rangle_{i<m})$. Then there exists $\mathcal{A}^*$ and $\mathcal{B}^*$ such that $\mathcal{A}^* \prec \mathcal{B}^*$, card $\mathcal{B}^* = \omega$, $R^{\mathcal{A}^*} = R^{\mathcal{B}^*}$, $\mathcal{A} \prec \mathcal{A}^*$, $\mathcal{B} \prec \mathcal{B}^*$, $b \in B^* - A^*$ and p is realized in $\mathcal{B}^*$.*

Proof. Let K be a 1-place relation symbol not occurring in $\mathscr{L}$. Let $\mathscr{L}_A^K$ be the language whose primitive symbols are those of $\mathscr{L}$ augmented by K and by names for the elements of A. Associated with each formula H of $\mathscr{L}_A^K$ is a formula H^K defined by induction on the length of H:

(i) if H has no quantifiers, then H^K is H;
(ii) if H is $(Ex)F$, then H^K is $(Ex)(K(x)\,\&\,F)$. H^K is called the restriction of H to K.

S is the following set of sentences:

(1) $T(\langle \mathcal{B}, b \rangle_{b \in B})$. $K(\underline{a})$ for all $a \in A$.
(2) H^K for every $H \in T(\langle \mathcal{A}, a \rangle_{a \in A})$.
(3) The universal closure of

$$[H^K(x_1, \ldots, x_n)\,\&\,K(x_1)\&\cdots\&K(x_n)] \to H(x_1, \ldots, x_n)$$

for every formula $H(x_1, \ldots, x_n)$ of $\mathscr{L}$.
(4) $(x)[R(x) \to K(x)]$.
(5) $\sim K(\underline{b})$, where $\underline{b}$ is the name of the given $b \in B - A$.
(6) $G(\underline{c})$ for every $G(x) \in p$, where $\underline{c}$ is an individual constant not occurring in (1)–(5). (For the sake of notational simplicity, let $n = 1$ and $m = 0$.)

S is consistent, since any finite subset of S can be satisfied by $\mathcal{B}$ if $K(x)$ is interpreted as "$x \in A$". Let $\mathcal{B}^*$ be a countable model of S. Let

$$A^* = \{b \mid \mathcal{B}^* \models K(\underline{b})\}.$$

For each n-place relation symbol J of $\mathscr{L}$, define

$$J^{\mathcal{A}^*} = J^{\mathcal{B}^*} \cap (A^*)^n.$$

Define the functions and distinguished elements of $\mathcal{A}^*$ similarly. By (1) $\mathcal{B} \prec \mathcal{B}^*$. By (1) and (2) $\mathcal{A} \prec \mathcal{A}^*$. By (3) $\mathcal{A}^* \prec \mathcal{B}^*$. By (4) $R^{\mathcal{A}^*} = R^{\mathcal{B}^*}$. By (5) $b \in B^* - A^*$. And by (6) p is realized in $\mathcal{B}^*$. $\qquad\square$

Proposition 22.2. *Suppose* $\mathcal{A} \prec \mathcal{B}$, *card* $\mathcal{B} = \omega$, $R^{\mathcal{A}} = R^{\mathcal{B}}$, $b \in B - A$ *and* $p \in S_n T(\langle \mathcal{A}, a_i \rangle_{i<m})$. *Then there exist* $\mathcal{A}^*$ *and* $\mathcal{B}^*$ *such that* $\mathcal{A}^* \prec \mathcal{B}^*$, *card* $\mathcal{B}^* = \omega$, $R^{\mathcal{A}^*} = R^{\mathcal{B}^*}$, $\mathcal{A} \prec \mathcal{A}^*$, $\mathcal{B} \prec \mathcal{B}^*$, $b \in B^* - A^*$ *and* p *is realized in* $\mathcal{A}^*$.

Proof. Same as that of 22.1. The only change is in clause (6): $G(\underline{c})$ is replaced by $G(\underline{c})$ & $K(\underline{c})$. □

Lemma 22.3. *Suppose* $\mathcal{A} \prec \mathcal{B}$, *card* $\mathcal{B} = \omega$, $R^{\mathcal{A}} = R^{\mathcal{B}}$ *and* $A \neq B$. *Then there exist* $\mathcal{A}^*$ *and* $\mathcal{B}^*$ *such that* $\mathcal{A} \prec \mathcal{A}^*$, $\mathcal{B} \prec \mathcal{B}^*$, $\mathcal{A}^* \prec \mathcal{B}^*$, $\mathcal{B}^*$ *is countable*, $R^{\mathcal{A}^*} = R^{\mathcal{B}^*}$, $A^* \neq B^*$, $\mathcal{A}^*$ *and* $\mathcal{B}^*$ *are homogeneous, and* $\mathcal{A}^* \approx \mathcal{B}^*$.

Proof. Similar to that of 20.5. By 22.1 and 22.2 there exists a diagram

$$\mathcal{B}_0 \prec \mathcal{B}_1 \prec \mathcal{B}_2 \prec \cdots$$
$$\curlyvee \quad \curlyvee \quad \curlyvee$$
$$\mathcal{A}_0 \prec \mathcal{A}_1 \prec \mathcal{A}_2 \prec \cdots$$

such that:

(1) $\mathcal{A} = \mathcal{A}_0$ and $\mathcal{B} = \mathcal{B}_0$.
(2) There is a b such that for all i, $b \in B_i - A_i$.
(3) $R^{\mathcal{A}_i} = R^{\mathcal{B}_i}$; $\mathcal{A}_i$ and $\mathcal{B}_i$ are countable.
(4) If $p \in S_n T$ is realized in $\mathcal{B}_{3i}$, then p is realized in $\mathcal{A}_{3i+1}$.
(5) Every finite partial automorphism of $\mathcal{B}_{3i+1}$ is immediately extendible in $\mathcal{B}_{3i+2}$ (cf. 20.5).
(6) Every finite partial automorphism of $\mathcal{A}_{3i+2}$ is immediately extendible in $\mathcal{A}_{3i+3}$.

Let $\mathcal{B}^* = \cup \{ \mathcal{B}_i \mid i < \omega \}$ and $\mathcal{A}^* = \cup \{ \mathcal{A}_i \mid i < \omega \}$. $\mathcal{A}^* \approx \mathcal{B}^*$ by 20.4. □

Lemma 22.4 (R. Vaught). *Suppose* $\mathcal{A} \prec \mathcal{B}$, $A \neq B$, *card* $\mathcal{B} = \omega$ *and* $R^{\mathcal{A}} = R^{\mathcal{B}}$. *Then there exists a* $\mathcal{C} \succ \mathcal{B}$ *such that card* $\mathcal{C} = \omega_1$ *and card* $R^{\mathcal{C}} \leq \omega$.

Proof. By 22.3 it is safe to assume $\mathcal{A}$ and $\mathcal{B}$ are homogenous and isomorphic. An elementary chain $\{\mathcal{C}_\delta \mid \delta < \omega_1\}$ is defined by induction on δ.

(i) $\mathcal{C}_0 = \mathcal{A}$.

(ii) Assume $\mathcal{C}_\delta \approx \mathcal{C}_0$. Choose $\mathcal{C}_{\delta+1}$ so that $\mathcal{C}_\delta \prec \mathcal{C}_{\delta+1}$, $\mathcal{C}_\delta \neq \mathcal{C}_{\delta+1}$, $R^{\mathcal{C}_\delta} = R^{\mathcal{C}_{\delta+1}}$ and $\mathcal{C}_\delta \approx \mathcal{C}_{\delta+1}$. $\mathcal{C}_{\delta+1}$ bears the same relation to $\mathcal{C}_\delta$ that $\mathcal{B}$ does to $\mathcal{A}$.

(iii) If λ is a limit ordinal, then $\mathcal{C}_\lambda = \cup\{\mathcal{C}_\delta \mid \delta < \lambda\}$. Assume $\mathcal{C}_\delta \approx \mathcal{C}_0$ for every $\delta < \lambda$. By 20.6 $\mathcal{C}_\lambda \approx \mathcal{C}_0$.

Let $\mathcal{C} = \cup\{\mathcal{C}_\delta \mid \delta < \omega_1\}$. $\qquad\qquad\qquad\qquad\qquad$ $\square$

$\langle \mathcal{A}, \mathcal{B} \rangle$ is a Vaughtian pair for T if there is a formula $R(x)$ such that $\mathcal{A} \models T$, $\mathcal{A} \prec \mathcal{B}$, $\mathcal{A} \neq \mathcal{B}$, $R^{\mathcal{A}} = R^{\mathcal{B}}$ and $R^{\mathcal{B}}$ is infinite.

Corollary 22.5. *If T is ω_1-categorical, then T has no Vaughtian pair.*

Proof. Let K be a 1-place relation symbol not occurring in T. For each formula H in the language of T, define H^K, the restriction of H to K, as in the proof of 22.1. S is the following set of sentences:

(1) T.

(2) The universal closure of

$$[H^K(x_1, \ldots, x_n) \,\&\, K(x_1) \,\&\, \cdots \,\&\, K(x_n)] \to H(x_1, \ldots, x_n)$$

for every formula $H(x_1, \ldots, x_n)$ in the language of T.

(3) $(x)[R(x) \to K(x)]$.

(4) $\sim K(\underline{b})$, where $\underline{b}$ is an individual constant not occurring in the language of T.

Suppose T has a Vaughtian pair $\langle \mathcal{C}, \mathcal{D} \rangle$ such that $R^{\mathcal{C}} = R^{\mathcal{D}}$ and $R^{\mathcal{D}}$ is infinite. Then S is consistent, since $\mathcal{D}$ can be constructed as a model of S by interpreting $K(x)$ as "$x \in \mathcal{C}$". Let $\mathcal{B}^*$ be a *countable* model of S. Define $\mathcal{A}^*$ as in the proof of 22.1.

Then $\langle \mathcal{A}^*, \mathcal{B}^* \rangle$ is a Vaughtian pair. By 22.4 T has a model $\mathcal{C}^*$ such that card $\mathcal{C}^* = \omega_1$ and card $R^{\mathcal{C}^*} = \omega$. By compactness T has a model $\mathcal{D}^*$ such that card $\mathcal{D}^* = \omega_1$ and card $R^{\mathcal{D}^*} = \omega_1$. Clearly $\mathcal{C}^* \not\approx \mathcal{D}^*$. □

Theorem 22.6 (R. Vaught). *If* $\kappa > \rho \geq \omega$, *then* $\langle \kappa, \rho \rangle \rightarrow \langle \omega_1, \omega \rangle$.

Proof. Let card $\mathcal{D} = \kappa$ and card $R^{\mathcal{D}} = \rho$. By the downward Skolem–Löwenheim theorem (11.2), there is a $\mathcal{C} \prec \mathcal{D}$ such that $R^{\mathcal{C}} = R^{\mathcal{D}}$ and card $\mathcal{C} = \rho$. Thus $\langle \mathcal{C}, \mathcal{D} \rangle$ is a Vaughtian pair for $T\mathcal{D}$. As in the proof of 22.5, $T\mathcal{D}$ has a model $\mathcal{D}^*$ such that card $\mathcal{D}^* = \omega_1$ and card $R^{\mathcal{D}^*} = \omega$. □

Section 23

Chang's Two-Cardinal Theorem

Let $\mathscr{L}$ be a countable language, and let $R(x)$ be a distinguished formula of $\mathscr{L}$. Suppose that $\mathcal{B}$ has the same similarity type as $\mathscr{L}$, and that card $\mathcal{B} >$ card $R^{\mathcal{B}} \geq \omega$. By 22.6 there is a $\mathcal{C} \equiv \mathcal{B}$ such that card $\mathcal{C} = \omega_1$ and card $R^{\mathcal{C}} = \omega$. Let κ be a regular uncountable cardinal such that $2^\rho \leq \kappa$ for all $\rho < \kappa$. The argument below furnishes a $\mathcal{C} \equiv \mathcal{B}$ such that card $\mathcal{C} = \kappa^+$ and card $R^{\mathcal{C}} = \kappa$. The assumptions on κ are needed to obtain certain saturated structures of cardinality κ, structures that will be utilized to omit a type as were the countable homogeneous structures in the proof of 22.4.

Proposition 23.1. *Suppose card $\mathcal{B} >$ card $R^{\mathcal{B}} \geq \omega$. Let κ be a regular uncountable cardinal such that $2^\rho \leq \kappa$ for every $\rho < \kappa$. Then there exist isomorphic saturated structures $\mathcal{A}_0$ and $\mathcal{A}_1$ such that $\mathcal{B} \equiv \mathcal{A}_0$, $\mathcal{A}_0 \prec \mathcal{A}_1$, $\mathcal{A}_0 \neq \mathcal{A}_1$, $R^{\mathcal{A}_0} = R^{\mathcal{A}_1}$ and card $\mathcal{A}_0 =$ card $\mathcal{A}_1 =$ card $R^{\mathcal{A}_1} = \kappa$.*

Proof. It follows from 11.2 (the downward Skolem–Löwenheim theorem) that there exists an $\mathcal{A} \prec \mathcal{B}$ such that $R^{\mathcal{A}} = R^{\mathcal{B}}$ and card $\mathcal{A} =$ card $R^{\mathcal{A}}$. Choose $b \in B - A$. Let K be a 1-place relation symbol not occurring in $\mathscr{L}$, and let $\mathscr{L}^K$ be $\mathscr{L}$ augmented by K. Define H^K as in the proof of 22.1, and let S

be the following set of sentences:

(1) $T\mathcal{B}$.
(2) The universal closure of

$$[K(x_1)\&\cdots\&K(x_n)\,\&\,H^K(x_1,\ldots,x_n)] \to H(x_1,\ldots,x_n)$$

for every formula $H(x_1,\ldots,x_n)$ of $\mathscr{L}$.
(3) $(x)[R(x) \to K(x)]$.
(4) $\sim K(\underline{b})$, where $\underline{b}$ names the chosen $b \in B - A$.

S is consistent, since $\mathcal{B}$ is a model of S when $K(x)$ is interpreted as "$x \in A$". Let $\mathcal{A}_1$ be a saturated model of S of cardinality κ. $\mathcal{A}_1$ exists by 16.5. Let

$$A_0 = \{a \mid a \in A_1 \,\&\, \mathcal{A}_1 \models K(\underline{a})\}.$$

The relations and functions of $\mathcal{A}_0$ are defined by restriction. Thus for each n-place relation symbol J of $\mathscr{L}$,

$$J^{\mathcal{A}_0} = J^{\mathcal{A}_1} \cap (A_0)^n.$$

By (1) $\mathcal{B} \equiv \mathcal{A}_1$. By (2) $\mathcal{A}_0 \prec \mathcal{A}_1$. By (3) $R^{\mathcal{A}_0} = R^{\mathcal{A}_1}$. And by (4) $\mathcal{A}_0 \neq \mathcal{A}_1$. (1) implies card $R^{\mathcal{A}_1} \geq \omega$. Since $\mathcal{A}_1$ is saturated, card $R^{\mathcal{A}_1} = \kappa$. To see that $\mathcal{A}_0$ is saturated, let $p \in S_1 T(\langle \mathcal{A}_0, x \rangle_{x \in X})$, where card $X < \kappa$. The conjunction of any finite subset of $p \cup \{K(x)\}$ can be satisfied in $\mathcal{A}_1$, hence p is realized in $\mathcal{A}_0$. $\mathcal{A}_0 \approx \mathcal{A}_1$ by 16.3. $\qquad\square$

Let $\mathcal{A}$ be such that card $R^{\mathcal{A}} \geq \omega$. $\mathcal{A}$ is said to be R-saturated if every p with the following properties is realized in $\mathcal{A}$:

$$p \in S_1 T(\langle \mathcal{A}, y \rangle_{y \in Y}), \quad \text{card } Y < \text{card } \mathcal{A} \text{ and } R(x) \in p.$$

R-saturated structures can be used to circumvent a shortcoming of saturated structures; namely, the union of an elementary chain of saturated structures need not be saturated. If the union always were, then Chang's two-cardinal theorem could be proved as follows. Start with the pair $\mathcal{A}_0$, $\mathcal{A}_1$ described in the conclusion of 23.1. Construct an elementary chain $\{\mathcal{A}_\delta \,|\, \delta < \kappa^+\}$ by induction on δ.

(i) Fix $\delta > 1$ and assume $\mathcal{A}_\gamma$ is isomorphic to $\mathcal{A}_0$, hence saturated, for all $\gamma < \delta$.

(ii) If δ is a limit ordinal, set $\mathcal{A}_\delta = \cup \{\mathcal{A}_\gamma \,|\, \gamma < \delta\}$. Then (if the union of an elementary chain of saturated structures were always saturated) $\mathcal{A}_\delta$ is saturated, hence isomorphic to $\mathcal{A}_0$.

(iii) If δ is a successor ordinal, choose $\mathcal{A}_\delta$ so that $\mathcal{A}_\delta$ bears the same relation to $\mathcal{A}_{\delta-1}$ that $\mathcal{A}_1$ bears to $\mathcal{A}_0$. Thus $\mathcal{A}_{\delta-1} \prec \mathcal{A}_\delta$, $\mathcal{A}_{\delta-1} \neq \mathcal{A}_\delta$, $\mathcal{A}_{\delta-1} \approx \mathcal{A}_\delta$ and $R^{\mathcal{A}_{\delta-1}} = R^{\mathcal{A}_\delta}$.

Let $\mathcal{C} = \cup \{\mathcal{A}_\delta \,|\, \delta < \kappa^+\}$. Then card $\mathcal{C} = \kappa^+$ and $R^{\mathcal{C}} = R^{\mathcal{A}_0}$. Although the parenthetical remark in step (ii) is false, it is possible to prove Chang's theorem by following the outline of (i), (ii) and (iii) with R-saturated structures in place of saturated ones. In order to convert step (ii) into something valid, a harmless assumption is made about $\mathcal{A}_0$. The language of $\mathcal{A}_0$ has a 2-place relation symbol $N(y, x)$ such that (1) and ($2n$) are valid in $\mathcal{A}_0$ for all $n > 0$.

(1) $N(y, x) \rightarrow R(y) \,\&\, R(x)$.

($2n$) $R(x_1) \,\&\, \cdots \,\&\, R(x_n)$
$\rightarrow (Ey)(x)[N(y, x) \leftrightarrow x = x_1 \vee \cdots \vee x = x_n]$.

If $\mathcal{A}_0$ has no such N, then one can be added. $\mathcal{A}_0$ was supplied by 23.1. The hypothesis of 23.1 concerned a $\mathcal{B}$ such that card $R^{\mathcal{B}} \geq \omega$. Adding a suitable $N^{\mathcal{B}}$ to $\mathcal{B}$ is no more difficult than choosing a one-to-one correspondence between $R^{\mathcal{B}}$ and the finite subsets of $R^{\mathcal{B}}$. The effect of adding $N^{\mathcal{B}}$ to $\mathcal{B}$ is to introduce "names" in $R^{\mathcal{B}}$ for the finite subsets of $R^{\mathcal{B}}$. Since $\mathcal{A}_0 \equiv \mathcal{B}$ and $\mathcal{A}_0$ is saturated, the effect of $N^{\mathcal{A}_0}$ is to provide "names" in $R^{\mathcal{A}_0}$ for subsets of $R^{\mathcal{A}_0}$ of cardinality less than card $\mathcal{A}_0$. $N^{\mathcal{A}_0}$ is said to be a naming relation for $\mathcal{A}_0$.

Lemma 23.2 (C. Chang). *Suppose* $\{\mathcal{A}_\delta \,|\, \delta < \lambda\}$ *is an elementary chain of R-saturated structures of cardinality* κ *such that* $R^{\mathcal{A}_\delta} = R^{\mathcal{A}_0}$ *for all* $\delta < \lambda$. *Assume* λ *is a limit,* $\lambda < \kappa^+$ *and* $\mathcal{A}_0$ *has a naming relation* $N^{\mathcal{A}_0}$. *Then*

$$\mathcal{A}_\lambda = \cup \{\mathcal{A}_\delta \,|\, \delta < \lambda\}$$

is R-saturated.

Proof. Let $p \in S_1 T(\langle \mathcal{A}_\lambda, y \rangle_{y \in Y})$ be such that card $Y <$ card $\mathcal{A}_\lambda$ and $R(x) \in p$. Thus

$$p = \{G_i(x) \,|\, i \in I\},$$

where card $I <$ card $\mathcal{A}_\lambda$. For each finite $J \subset I$, choose $a_J \in R^{\mathcal{A}_\lambda} = R^{\mathcal{A}_0}$ so that

$$\mathcal{A}_\lambda \models G_i(\underline{a}_J) \quad \text{for all } i \in J.$$

Fix $i \in I$; every member of A_λ mentioned in $G_i(x)$ occurs in $\mathcal{A}_{\delta_i}$ for some $\delta_i < \lambda$ such that card $I <$ card $\mathcal{A}_{\delta_i}$. Some $b_i \in R^{\mathcal{A}_0}$ is needed with the property that:

(a) $\mathcal{A}_{\delta_i} \models (x)[N(\underline{b}_i, x) \to G_i(x)]$;
(b) $\mathcal{A}_{\delta_i} \models N(\underline{b}_i, \underline{a}_J)$ for all J such that $i \in J$.

b_i will serve as a "name" for a subset of

$$\{a \,|\, \mathcal{A}_{\delta_i} \models G_i(\underline{a})\}$$

that includes all the a_J's such that $i \in J$. Since all the a_J's belong to $R^{\mathcal{A}_0}$, every finite set of a_J's that satisfy $G_i(x)$ has a "name" in $R^{\mathcal{A}_0}$. Consequently, the existence of b_i follows from the R-saturation of $\mathcal{A}_{\delta_i}$.

Choose $q \in S_1 T(\langle \mathcal{A}_0, b_i \rangle_{i \in I})$ so that

$$q \supset \{N(\underline{b}_i, x) \,|\, i \in I\}.$$

q exists because for each finite $J \subset I$, the conjunction of $\{N(\underline{b}_i, x) \,|\, i \in J\}$ is satisfiable by a_J. Any realization of q in $\mathcal{A}_0$

is a realization of p in $\mathcal{A}_\lambda$. But q is realized in $\mathcal{A}_0$ since $\mathcal{A}_0$ is R-saturated. $\square$

Lemma 23.3. *Suppose $\mathcal{A} \equiv \mathcal{B}$, card $\mathcal{A} =$ card $\mathcal{B}$, $\mathcal{B}$ is saturated, and $\mathcal{A}$ is R-saturated. Then there exists an $f\colon \mathcal{A} \overset{\equiv}{\to} \mathcal{B}$ such that $f[R^\mathcal{A}] = R^\mathcal{B}$.*

Proof. Let $A = \{a_\delta \,|\, \delta < \kappa\}$ and $R^\mathcal{B} = \{b_\delta \,|\, \delta < \text{card } R^\mathcal{B}\}$. A set $\{\langle a_\delta^*, b_\delta^* \rangle \,|\, \delta < \kappa\}$ is defined by induction on δ. Fix δ and assume

$$\langle \mathcal{A}, a_\gamma^* \rangle_{\gamma < \delta} \equiv \langle \mathcal{B}, b_\gamma^* \rangle_{\gamma < \delta}.$$

Case 1. δ is even. Let a_δ^* be the member of $A - \{a_\gamma^* \,|\, \gamma < \delta\}$ with least subscript. Let p be

$$\{F(x) \,|\, \langle \mathcal{A}, a_\gamma^* \rangle_{\gamma < \delta} \models F(a_\delta^*)\}.$$

Since $\mathcal{B}$ is saturated, p is realized in $\mathcal{B}$ by some b; set $b_\delta^* = b$.

Case 2. δ is odd and $\delta < \text{card } R^\mathcal{B}$. Let b_δ^* be the member of $R^\mathcal{B} - \{b_\gamma^* \,|\, \gamma < \delta\}$ with least subscript. Let q be

$$\{F(x) \,|\, \langle \mathcal{B}, b_\gamma^* \rangle_{\gamma < \delta} \models F(b_\delta^*)\}.$$

Clearly $R(x) \in q$. Since $\mathcal{A}$ is R-saturated, q is realized in $\mathcal{A}$ by some $a \in R^\mathcal{A}$; set $a_\delta^* = a$.

Define $f\colon A \to B$ by $f a_\delta^* = b_\delta^*$. By case 1 the domain of f is A. By case 2 f maps $R^\mathcal{A}$ onto $R^\mathcal{B}$. $\square$

Theorem 23.4 (C. Chang). *If $\mu' > \mu \geq \omega$ and κ is a regular uncountable cardinal such that $2^\rho \leq \kappa$ for every $\rho < \kappa$, then*

$$\langle \mu', \mu \rangle \to \langle \kappa^+, \kappa \rangle.$$

Proof. Let $\mathcal{B}$ have two-cardinal type $\langle \mu', \mu \rangle$. Assume $\mathcal{B}$ has a naming relation $N^\mathcal{B}$. By 23.1 there exist saturated structures $\mathcal{A}_0$ and $\mathcal{A}_1$ of cardinality κ such that $\mathcal{B} \equiv \mathcal{A}_0$, $\mathcal{A}_0 \prec \mathcal{A}_1$, $\mathcal{A}_0 \neq \mathcal{A}_1$ and $R^{\mathcal{A}_0} = R^{\mathcal{A}_1}$. An elementary chain $\{\mathcal{A}_\delta \,|\, \delta < \kappa^+\}$ is defined by recursion on δ.

(i) Fix $\delta > 1$ and assume $\mathcal{A}_\gamma$ is R-saturated, card $\mathcal{A}_\gamma = \kappa$ and $R^{\mathcal{A}_\gamma} = R^{\mathcal{A}_0}$ for all $\gamma < \delta$.

(ii) If δ is a limit ordinal, set $\mathcal{A}_\delta = \cup\{\mathcal{A}_\gamma \mid \gamma < \delta\}$. Then $\mathcal{A}_\delta$ is R-saturated by 23.2.

(iii) If δ is a successor ordinal, then by 23.3, there is an f: $\mathcal{A}_{\delta-1} \overset{\equiv}{\to} \mathcal{A}_0$ such that $f[R^{\mathcal{A}_{\delta-1}}] - R^{\mathcal{A}_0}$. Choose $\mathcal{A}_\delta$ so that $\mathcal{A}_\delta$ bears the same relation to $\mathcal{A}_{\delta-1}$ that $\mathcal{A}_1$ bears to $f[\mathcal{A}_{\delta-1}]$. Thus $\mathcal{A}_\delta \succ \mathcal{A}_{\delta-1}$, $\mathcal{A}_\delta \neq \mathcal{A}_{\delta-1}$, $R^{\mathcal{A}_\delta} = R^{\mathcal{A}_{\delta-1}}$ and $\mathcal{A}_\delta$ is R-saturated.

Let $\mathcal{C} = \cup\{\mathcal{A}_\delta \mid \delta < \kappa^+\}$. Then card $\mathcal{C} = \kappa^+$ and $R^{\mathcal{C}} = R^{\mathcal{A}_0}$.

$\square$

A parting remark on the role of saturated structures in the proof of Chang's two-cardinal theorem: The R-saturation of $\mathcal{A}_0$ suggests that every possible future addition to $R^{\mathcal{A}_0}$ has already been realized in $\mathcal{A}_0$. Thus the richness of $R^{\mathcal{A}_0}$ makes it plausible that $\mathcal{A}_0$ can be enlarged considerably without realizing the type: "x is a new member of R."

R. B. Jensen has shown: if Gödel's axiom of constructibility $(V = L)$ holds, then

$$\langle \mu', \mu \rangle \to \langle \kappa^+, \kappa \rangle$$

whenever $\mu' > \mu \geq \omega$ and $\kappa \geq \omega$. His proof, when κ is singular, utilizes the ideas above.

Section 24

Keisler's Two-Cardinal Theorem

Yet another way of omitting a type is by means of the completeness theorem (24.2) for ω-logic.

Let τ be a similarity type. Conjoined with τ is the first order language $\mathcal{L}_\tau$ defined in Sec. 4. Also concomitant with τ is the ω-logic language $\mathcal{L}_\tau^\omega$, whose primitive symbols are those of $\mathcal{L}_\tau$ augmented by a 1-place relation symbol N and a set $\{\underline{n} \mid n < \omega\}$ of individual constants. It is assumed that neither N nor any member of $\{\underline{n} \mid n < \omega\}$ occur in $\mathcal{L}_\tau$. The terms and formulas of $\mathcal{L}_\tau^\omega$ are generated from the primitive symbols of $\mathcal{L}_\tau^\omega$ as in Sec. 4.

$\mathcal{A}$ is an ω-structure if $\{n \mid n < \omega\} \subset A$. If $\mathcal{A}$ is an ω-structure of type τ, then $\mathcal{A}$ can be expanded to $\mathcal{A}^\omega$, a structure in which the sentences of $\mathcal{L}_\tau^\omega$ have definite truth values. $\mathcal{A}^\omega$ is $\mathcal{A}$ augmented by the 1-place relation

$$N^{\mathcal{A}} = \{n \mid n < \omega\}$$

and the set $\{n \mid n < \omega\}$ of distinguished elements. Let $H(x_1, \ldots, x_n)$ be a formula of $\mathcal{L}_\tau^\omega$, and let $a_1, \ldots, a_n \in A$. Then $\mathcal{A} \models H(\underline{a}_1, \ldots, \underline{a}_n)$ if $\mathcal{A}^\omega \models H(\underline{a}_1, \ldots, \underline{a}_n)$ in the sense of Sec. 5 with $\underline{n}$ interpreted as n and $N(x)$ as $x \in N^{\mathcal{A}}$.

$\mathcal{L}_\tau^\omega$ has considerably more expressive power than $\mathcal{L}_\tau$. For example every model of $(x)(Ey)[x = y \,\&\, N(y)]$ is countable.

Let S be a set of sentences of $\mathcal{L}_\tau^\omega$ that includes $\{\underline{m} \neq \underline{n} \,|\, m \neq n\}$. S is ω-consistent if S is consistent (in the sense of Sec. 7), and if for every sentence of $\mathcal{L}_\tau^\omega$ of the form $(Ex)[N(x) \,\&\, F(x)]$, the consistency of

$$S \cup \{(Ex)[N(x) \,\&\, F(x)]\}$$

implies the consistency of $S \cup \{F(\underline{n})\}$ for some n. The notion of ω-consistency is not finitary; i.e. it can happen that every finite $S_0 \subset S$ is ω-consistent but S is not. Nonetheless the notion of ω-consistency is absolute.

Proposition 24.1. *Let S be an ω-consistent set of sentences of ω-logic, and let G be a sentence in the language of S. If $S \cup \{G\}$ is consistent, then $S \cup \{G\}$ is ω-consistent.*

Proof. Suppose $S \cup \{G, (Ex)[N(x) \,\&\, F(x)]\}$ is consistent. Then $S \cup \{(Ex)[G \,\&\, N(x) \,\&\, F(x)]\}$ is consistent, since G is a sentence. But then for some n,

$$S \cup \{G \,\&\, F(\underline{n})\}$$

is consistent. $\qquad\qquad\qquad\qquad\qquad\qquad\qquad\qquad\qquad\qquad\square$

Theorem 24.2 (S. Orey). *If S is a countable, ω-consistent set of sentences of ω-logic, then S has a model. (cf. Exercise 24.7.)*

Proof. Similar to the Henkin construction of 7.1 save for an essential use of the countability of S. Let $\{\underline{c_i} \,|\, i < \omega\}$ be a sequence of individual constants not occurring in S. Let $\{F_i(x) \,|\, i < \omega\}$ be an enumeration of all formulas (in the language of S increased by the $\underline{c_i}$'s) whose sole free variable, if any, is x. A sequence $\{T_i \,|\, i < \omega\}$ of consistent sets of sentences is defined by induction on i. Measures will be taken to insure that $T_\omega = \cup \{T_i \,|\, i < \omega\}$ is ω-consistent and that every sentence or its negation belongs to T_ω.

Let $T_0 = S$. Fix $i \geq 0$. Assume that T_i has been defined so that T_i is consistent and is the result of adding finitely many sentences to S.

Case 1. x is a free variable of $F_i(x)$. Choose a $\underline{c_k}$ that occurs neither in T_i nor in $F_i(x)$. Let

$$T_{i+1} = T_i \cup \{(Ex)F_i(x) \rightarrow F_i(\underline{c_k})\}.$$

T_{i+1} is consistent because T_i is; the argument is the same as that of 7.1.

Case 2. $F_i(x)$ is a sentence, call it F_i.

Case 2a. F_i is not of the form $(Ex)[N(x) \,\&\, H(x)]$. If $T_i \cup \{F_i\}$ is consistent, let $T_{i+1} = T_i \cup \{F_i\}$; if not, let $T_{i+1} = T_i \cup \{\sim F_i\}$. As in 7.1 T_{i+1} is consistent.

Case 2b. F_i is of the form $(Ex)[N(x) \,\&\, H(x)]$.

Case 2b(i). $T_i \cup \{(Ex)[N(x) \,\&\, H(x)]\}$ is not consistent. Let $T_{i+1} = T_i \cup \{\sim F_i\}$.

Case 2b(ii). $T_i \cup \{(Ex)[N(x) \,\&\, H(x)]\}$ is consistent. By 24.1 there is an n such that

$$T_i \cup \{N(\underline{n}) \,\&\, H(\underline{n})\}$$

is consistent, since T_i differs finitely from S. Let T_{i+1} be

$$T_i \cup \{F_i, N(\underline{n}), H(\underline{n})\}.$$

Let $T_\omega = \cup \{T_i \mid i < \omega\}$. By case 2 every sentence or its negation belongs to T_ω. T_ω is consistent since every T_i is. To see that T_ω is ω-consistent, suppose $T_\omega \cup \{(Ex)[N(x) \,\&\, H(x)]\}$ is consistent. It follows from case 2b that

$$T_\omega \cup \{N(\underline{n}) \,\&\, H(\underline{n})\}$$

is consistent for some n. Now a model $\mathcal{A}$ of S can be constructed as in 7.1 from T_ω. The members of A are equivalence classes of constant terms occurring in T. The ω-consistency of T_ω is needed to show that $\mathcal{A} \models F$ iff $F \in T_\omega$, where F is any sentence in the ω-logic language of T_ω. $\qquad\square$

Let $\mathcal{B}$ be a structure whose underlying language includes a 1-place relation R and a 2-place relation $\leq$. Suppose $\leq^{\mathcal{B}}$ is a linear ordering of B with no greatest member. Let $F(y, z)$ be a formula in the language of $\langle \mathcal{B}, b \rangle_{b \in B}$. $\mathcal{B}$ is said to have property $K_{F(y,z)}$ if $\mathcal{B} \models$ (i) $\rightarrow$ (ii), where (i) is

$$(u)(Ey)(Ez)[u \leq y \ \& \ R(z) \ \& \ F(y, z)],$$

and (ii) is

$$(Ez)(u)(Ey)[u \leq y \ \& \ R(z) \ \& \ F(y, z)].$$

$\mathcal{B}$ has property K if $\mathcal{B}$ has property $K_{F(y,z)}$ for every $F(y, z)$ in the language of $\langle \mathcal{B}, b \rangle_{b \in B}$. An intuitive version of K is: if f is a Skolem function of $\mathcal{B}$ such that $fy \in R^{\mathcal{B}}$ for arbitrarily large y, then there exists a $z \in R^{\mathcal{B}}$ such that $fy = z$ for arbitrarily large y.

Lemma 24.3 (H. J. Keisler). *Suppose $\mathcal{B}$ is countable and has property K. Then there exist a countable $\mathcal{C} \succ \mathcal{B}$ such that $C \neq B$ and $R^{\mathcal{C}} = R$.*

Proof. If $R^{\mathcal{B}}$ is finite, then the lemma follows from 7.3. Suppose $R^{\mathcal{B}}$ is infinite and

$$\{b \,|\, b \in R^{\mathcal{B}}\} = \{n \,|\, n < \omega\}.$$

Let S be the following collection of sentences of ω-logic:

(i) $T(\langle \mathcal{B}, b \rangle_{b \in B})$;

(ii) $\{\underline{b} < \underline{c} \,|\, b \in B\}$, where $\underline{c}$ is an individual constant not occurring in (i);

(iii) $(x)[R(x) \leftrightarrow N(x)]$.

Any countable model of S will do for $\mathcal{C}$, so by 24.2 it suffices to check that S is ω-consistent. Every finite subset of S is satisfiable in $\mathcal{B}$, since $\leq^{\mathcal{B}}$ has no greatest member; hence S is consistent. Let $F(y,z)$ be a formula in the language of $\langle \mathcal{B}, b \rangle_{b \in B}$ such that

$$S \cup \{(Ez)[R(z) \; \& \; F(\underline{c}, z)]\}$$

is consistent. Since S includes $T(\langle \mathcal{B}, b \rangle_{b \in B})$ and c is intuitively equivalent to an arbitrarily large member of B, it follows that

$$(u)(Ey)(Ez)[u \leq y \; \& \; R(z) \; \& \; F(y, z)]$$

is true in $\mathcal{B}$. Since $\mathcal{B}$ has property $K_{F(y,z)}$, there is an n such that

$$(u)(Ey)[u \leq y \; \& \; R(\underline{n}) \; \& \; F(y, \underline{n})]$$

is true in $\mathcal{B}$. But then

$$S \cup \{R(\underline{n}) \; \& \; F(\underline{c}, \underline{n})\}$$

is consistent, because $\leq^{\mathcal{B}}$ has no greatest member. $\qquad\square$

Theorem 24.4 (H. J. Keisler). *Suppose* card $\mathcal{A} >$ card $R^{\mathcal{A}}$ $\geq \omega$. *Then there exist* $\mathcal{B}$ *and* $\mathcal{C}$ *such that* $\mathcal{B} \prec \mathcal{A}$, $\mathcal{B} \prec \mathcal{C}$, card $\mathcal{B} = \omega$, card $\mathcal{C} = \omega_1$ *and* $R^{\mathcal{B}} = R^{\mathcal{C}}$.

Proof. By the downward Skolem–Löwenheim theorem (11.2), it is safe to assume

$$\text{card } \mathcal{A} = (\text{card } R^{\mathcal{A}})^+.$$

Add a 2-place relation symbol $\leq$ to the language of $\mathcal{A}$ and a relation $\leq^{\mathcal{A}} A$ so that $\leq^{\mathcal{A}}$ is a wellordering of A isomorphic to the wellordering of the ordinals less than card $\mathcal{A}$. Since card $\mathcal{A}$ is regular and greater than card $R^{\mathcal{A}}$, it follows that $\mathcal{A}$ has property K.

Choose a countable $\mathcal{B} \prec \mathcal{A}$. Clearly $\mathcal{B}$ has property K. An elementary chain $\{\mathcal{B}_\delta \mid \delta < \omega_1\}$ is defined by induction on δ:

(i) $\mathcal{B}_0 = \mathcal{B}$.

(ii) If δ is a limit ordinal, then $\mathcal{B}_\delta = \cup \{\mathcal{B}_\alpha \mid \alpha < \delta\}$. $\mathcal{B}_\delta$ has property K since $\mathcal{B}_0$ has it and $\mathcal{B}_0 \prec \mathcal{B}_\delta$.

(iii) Assume $\mathcal{B}_\delta$ has property K. Then by 24.3 there is a $\mathcal{B}_{\delta+1} \succ \mathcal{B}_\delta$ such that $B_{\delta+1} \neq B_\delta$ and $R^{\mathcal{B}_{\delta+1}} = R^{\mathcal{B}_\delta}$.
Let $\mathcal{C} = \cup \{\mathcal{B}_\delta \mid \delta < \omega_1\}$. $\qquad\qquad\qquad\square$

The proof of 24.4 reveals some of the hidden power of first order expressibility. $\mathcal{A}$ has property K because: card $\mathcal{A}$ is regular, card $\mathcal{A} >$ card $R^{\mathcal{A}}$ and $\leq^{\mathcal{A}}$ is a wellordering of A isomorphic to card $\mathcal{A}$. Property K is first order; i.e. if $\mathcal{B} \equiv \mathcal{C}$ and $\mathcal{B}$ has property K, then $\mathcal{C}$ has property K. But the notion of wellordering is not first order. (There exist linear orderings $\mathcal{A}$ and $\mathcal{B}$ such that $\mathcal{A} \equiv \mathcal{B}$, $\mathcal{A}$ is a wellordering and $\mathcal{B}$ is not.) Thus K is only a first order remnant of a second order property of $\mathcal{A}$, but it suffices for the type omitting argument (24.3) needed in 24.4.

The next result is an archetypic example of the use of a two-cardinal theorem to omit a type.

Corollary 24.5 (H. J. Keisler). *Let T be a countable theory such that every model of T of cardinality ω_1 is homogeneous. Then every uncountable model of T is homogeneous.*

Proof. Let $\mathcal{A}$ be an uncountable model of T. Suppose there is an elementary partial automorphism g of $\mathcal{A}$, of lesser cardinality than $\mathcal{A}$, which is not immediately extendible. Thus

$$\langle \mathcal{A}, u \rangle_{u \in U} \equiv \langle \mathcal{A}, gu \rangle_{u \in U},$$

but for some $c \in A$

$$\langle \mathcal{A}, u, c \rangle_{u \in U} \not\equiv \langle \mathcal{A}, gu, b \rangle_{u \in U}$$

for any $b \in A$. The addition of some relations and functions to $\mathcal{A}$ makes it possible to express the above failure of homogeneity by a single sentence of ω-logic. Let $G^{\mathcal{A}}$ be a 2-place relation such that

$$\langle a_1, a_2 \rangle \in G^{\mathcal{A}} \leftrightarrow ga_1 = a_2.$$

Assume $\{n \mid n < \omega\} \subset A$. Let $t^{\mathcal{A}}$ be a 3-place function such that for every $\langle a_1, \dots, a_n \rangle \in A^n$, there is an $a \in A$ with the property that

$$t^{\mathcal{A}}(a, n, i) = a_i$$

whenever $l \leq l \leq n$. $t^{\mathcal{A}}$ allows quantification over finite sequences of members of A. Thus

$$(Ex_1) \cdots (Ex_n) Q(x_1, \dots, x_n)$$

can be replaced by

$$(Ex) Q(t(x, \underline{n}, \underline{1}), \dots, t(x, \underline{n}, \underline{n})).$$

Let $S^{\mathcal{A}}$ be a 3-place relation such that for all $n, e < \omega$: $\langle n, e, a \rangle \in S^{\mathcal{A}}$ iff e is a Gödel number of the formula $F(x_1, \dots, x_n)$ and

$$\mathcal{A} \models F(t(\underline{a}, \underline{n}, \underline{1}), \dots, t(\underline{a}, \underline{n}, \underline{n})).$$

Let $\mathcal{A}^*$ be $\mathcal{A}$ augmented by $G^{\mathcal{A}}$, $t^{\mathcal{A}}$, and $S^{\mathcal{A}}$. Then the failure of g to be immediately extendible in $\mathcal{A}$ can be expressed by a single sentence of ω-logic true in $\mathcal{A}^*$

By 24.4 there exist $\mathcal{B}^*$ and $\mathcal{C}^*$ such that $\mathcal{B}^* \prec \mathcal{A}^*$, $\mathcal{B}^* \prec \mathcal{C}^*$, $G^{\mathcal{B}^*} = G^{\mathcal{C}^*}$, card $\mathcal{B}^* = \omega$ and card $\mathcal{C}^* = \omega_1$. ($t^{\mathcal{A}}$ allows $G^{\mathcal{A}}$ to be construed as a 1-place relation.) Then $G^{\mathcal{C}^*}$ is the graph of a countable, elementary partial automorphism of $\mathcal{C}^*$, not immediately extendible in $\mathcal{C}^*$, so $\mathcal{C}^*$ is a nonhomogeneous model of T of cardinality ω_1. $\qquad \square$

S. Shelah has strengthened the conclusion of 24.5 to: every uncountable model of T is saturated.

Exercise 24.6. Suppose T is a countable theory with the property that every model of T of cardinality ω_1 is saturated. Show every uncountable model of T is saturated.

Exercise 24.7. Let S be a countable set of sentences of ω-logic. Suppose S has a model. Is S ω-consistent?

Section 25

Categories and Functors

The Morley analysis of 1-types begins in Sec. 27. Its essential properties can be clearly stated in the language of categories and functors; so it is not surprising that similar analyses ([Sh1], [Sh2]) of 1-types have similar properties. The Morley derivative will be presented in Sec. 29 as an operation that acts on certain contravariant functors. The present section reviews all the category theoretic notions needed in Sec. 29.

A category $\mathcal{K}$ consists of objects $\mathcal{A}, \mathcal{B}, \mathcal{C}, \ldots$; and maps $f\colon \mathcal{A} \to \mathcal{B}$, $g\colon \mathcal{B} \to \mathcal{C}, \ldots$. Associated with each pair

$$f\colon \mathcal{A} \to \mathcal{B}, \quad g\colon \mathcal{B} \to \mathcal{C}$$

of maps is a map $gf\colon \mathcal{A} \to \mathcal{C}$ called the composite of f and g. Associated with each object $\mathcal{A}$ is a map $1_{\mathcal{A}}\colon \mathcal{A} \to \mathcal{A}$ called the identity on $\mathcal{A}$. The composite and the identity obey two axioms:

(i) $h(gf) = (hg)f$, where $f\colon \mathcal{A} \to \mathcal{B}$, $g\colon \mathcal{B} \to \mathcal{C}$ and $h\colon \mathcal{C} \to \mathcal{D}$.
(ii) $1_{\mathcal{B}}f = f$ and $g1_{\mathcal{B}} = g$, where $f\colon \mathcal{A} \to \mathcal{B}$ and $g\colon \mathcal{B} \to \mathcal{C}$.

$\mathcal{K}$ admits filtrations if for each pair $\mathcal{A}$, $\mathcal{B}$ of objects of $\mathcal{K}$, there exist an object $\mathcal{C}$ and maps $f\colon \mathcal{A} \to \mathcal{C}$ and $g\colon \mathcal{B} \to \mathcal{C}$.

$\mathcal{K}$ admits filtrations with amalgamation if for each pair $j\colon \mathcal{D} \to \mathcal{A}$, $k\colon \mathcal{D} \to \mathcal{B}$ of maps of $\mathcal{K}$, there exist an object $\mathcal{C}$ and maps $f\colon \mathcal{A} \to \mathcal{C}$ and $g\colon \mathcal{B} \to \mathcal{C}$ such that $fj = gk$.

Directed sets were defined in Sec. 10. A direct system $\{\mathcal{A}_i, f_{ij}\}$ in $\mathcal{K}$ consists of a directed set $\langle D, \leq \rangle$, a family $\{\mathcal{A}_i \mid i \in D\}$ of objects of $\mathcal{K}$, and a family $\{f_{ij}\colon \mathcal{A}_i \to \mathcal{A}_j \mid i \leq j\}$ of maps of $\mathcal{K}$ such that:

(a) $f_{ii} = 1_{\mathcal{A}_i}$,

(b) $f_{ik} = f_{jk}f_{ij}$ whenever $i \leq j \leq k$.

   A direct limit in $\mathcal{K}$ of the direct system $\{\mathcal{A}_i, f_{ij}\}$ consists of an object $\mathcal{A}$ of $\mathcal{K}$ and a family $\{f_i \colon \mathcal{A}_i \to \mathcal{A} \mid i \in D\}$ of maps of $\mathcal{K}$ such that:

(c) $f_j f_{ij} = f_i$ whenever $i \leq j$.

(d) Let $\mathcal{B}$ be any object of $\mathcal{K}$ and $\{g_i \colon \mathcal{A}_i \to \mathcal{B} \mid i \in D\}$ be any family of maps of $\mathcal{K}$ such that $g_j f_{ij} = g_i$ whenever $i \leq j$. Then there exists a unique $g \colon \mathcal{A} \to \mathcal{B}$ in $\mathcal{K}$ such that $g f_i = g_i$ for all i.

It follows from (d), the universal property of the direct limit, that all direct limits in $\mathcal{K}$ of $\{\mathcal{A}_i, f_{ij}\}$, if there are any, are isomorphic. The direct limit of $\{\mathcal{A}_i, f_{ij}\}$ is denoted by $\varinjlim \mathcal{A}_i$ or by $\mathcal{A}_\infty$. $\mathcal{K}$ admits direct limits if every direct system in $\mathcal{K}$ has a direct limit in $\mathcal{K}$. Theorem 10.1 and Proposition 10.2 say: the category of all structures and all elementary monomorphisms admits direct limits.

An inverse system in $\mathcal{K}$ consists of a directed set $\langle D, \leq \rangle$, a family $\{\mathcal{A}_i \mid i \in D\}$ of objects of $\mathcal{K}$, and a family $\{f_{ji} \colon \mathcal{A}_j \to \mathcal{A}_i \mid i \leq j\}$ of maps of $\mathcal{K}$ such that:

(1) $f_{ii} = 1_{\mathcal{A}_i}$,

(2) $f_{ki} = f_{ji}f_{kj}$ whenever $i \leq j \leq k$.

   An inverse limit in $\mathcal{K}$ of the inverse system $\{\mathcal{A}_i, f_{ji}\}$ consists of an object $\mathcal{A}$ of $\mathcal{K}$ and a family $\{f_i \colon \mathcal{A} \to \mathcal{A}_i \mid i \in D\}$ of maps of $\mathcal{K}$ such that

(3) $f_{ji}f_j = f_i$ whenever $i \leq j$.

(4) Let $\mathcal{B}$ be any object of $\mathcal{K}$ and $\{g_i \colon \mathcal{B} \to \mathcal{A}_i \mid i \in D\}$ be any family of maps of $\mathcal{K}$ such that $f_{ji}g_j = g_i$ whenever $i \leq j$. Then there exists a unique $g \colon \mathcal{B} \to \mathcal{A}$ in $\mathcal{K}$ such that $f_i g = g_i$ for all i.

It follows from (4), the universal property of the inverse limit, that all inverse limits in $\mathcal{K}$ of $\{\mathcal{A}_i, f_{ji}\}$, if there are any, are isomorphic. The inverse limit of $\{\mathcal{A}_i, f_{ji}\}$ is denoted by $\varprojlim \mathcal{A}_i$ or by $\mathcal{A}_\infty$. $\mathcal{K}$ admits inverse limits if every inverse system in $\mathcal{K}$ has an inverse limit in $\mathcal{K}$. It will be shown in Sec. 26 that $\mathcal{H}$, the category of compact Hausdorff spaces and continuous onto maps, admits inverse limits.

Let $\mathcal{K}_1$ and $\mathcal{K}_2$ be categories. A contravariant functor F: $\mathcal{K}_1 \to \mathcal{K}_2$ assigns to each object $\mathcal{A}$ of $\mathcal{K}_1$, an object $F\mathcal{A}$ of $\mathcal{K}_2$, and to each map $f: \mathcal{A} \to \mathcal{B}$ of $\mathcal{K}_1$, a map $Ff: F\mathcal{B} \to F\mathcal{A}$ of $\mathcal{K}_2$ so that:

(i) $F1_\mathcal{A} = 1_{F\mathcal{A}}$,

(ii) $Fgf = FfFg$, where $f: \mathcal{A} \to \mathcal{B}$ and $g: \mathcal{B} \to \mathcal{C}$.

If $\{\mathcal{A}_i, f_{ij}\}$ is a direct system of $\mathcal{K}_1$, then $\{F\mathcal{A}_i, Ff_{ij}\}$ is an inverse system of $\mathcal{K}_2$. F preserves limits if:

(a) $\{F\mathcal{A}_i, Ff_{ij}\}$ has an inverse limit in $\mathcal{K}_2$ whenever $\{\mathcal{A}_i, f_{ij}\}$ has a direct limit in $\mathcal{K}_1$.

(b) Suppose the direct limit of $\{\mathcal{A}_i, f_{ij}\}$ consists of $\mathcal{A}_\infty$ and $\{f_i: \mathcal{A}_i \to \mathcal{A}_\infty\}$, and the inverse limit of $\{F\mathcal{A}_i, Ff_{ij}\}$ consists of $\varprojlim F\mathcal{A}_i$ and $\{g_i: \varprojlim F\mathcal{A}_i \to F\mathcal{A}_i\}$. Let

$$g: F\mathcal{A}_\infty \to \varprojlim F\mathcal{A}_i,$$

be the unique map such that $g_i g = Ff_i$ for all i. Then g is an isomorphism.

$\mathcal{K}^*$ is a subcategory of $\mathcal{K}$ if:

(a) Every object of $\mathcal{K}^*$ is an object of $\mathcal{K}$.

(b) Every map of $\mathcal{K}^*$ is a map of $\mathcal{K}$.

(c) If $f: \mathcal{A} \to \mathcal{B}$ and $g: \mathcal{B} \to \mathcal{C}$ belong to $\mathcal{K}^*$, then the composite of f and g in $\mathcal{K}^*$ equals the composite of f and g in $\mathcal{K}$.

$\mathcal{K}^*$ is a full subcategory of $\mathcal{K}$ if $g: \mathcal{A} \to \mathcal{B}$ belongs to $\mathcal{K}^*$ whenever $\mathcal{A}$ and $\mathcal{B}$ belong to $\mathcal{K}^*$ and g belongs to $\mathcal{K}$.

Section 26

Inverse Systems of
Compact Hausdorff Spaces

The Morley analysis of 1-types is centered on the notion of inverse limit. The key property of the Morley derivative, to be established in Sec. 29, is its commutativity with the inverse limit operation.

Let $\mathcal{H}$ be the category of all compact Hausdorff spaces and all continuous onto maps. Suppose $f\colon \mathcal{A} \to \mathcal{B}$ is a map. If $x \in B$, then $f^{-1}x$ is $\{a \mid a \in A \,\&\, fa = x\}$. If $U \subset B$, then $f^{-1}[U]$ is $\{a \mid a \in A \,\&\, fa \in U\}$.

Proposition 26.1. $\mathcal{H}$ *admits inverse limits.*

Proof. Suppose $\{X_i, f_{ji}\}$ is an inverse system in $\mathcal{H}$ directed by $\langle D, \leq \rangle$. Let X_∞ be $\{x \mid x \in \prod_{i \in D} X_i$ and $f_{ji}x_j = x_i$ for all $j \geq i\}$, where x_i is the ith coordinate of x. Assign the product topology to $\prod_i X_i$. Then X_∞ is a closed, hence compact subset of $\prod_i X_i$. Give X_∞ the subspace topology. Define

$$f_{\infty i}\colon X_\infty \to X_i$$

by $f_{\infty i}x = x_i$. Clearly $f_{\infty i}$ is continuous. The fact that $f_{\infty i}$ is onto follows from the compactness of $\prod_i X_i$. A typical basic open subset of X_∞ is $f_{\infty j}^{-1}[U]$, where U is a basic open subset of X_j.

To see that $\{X_\infty, f_{\infty i}\}$ is the inverse limit of $\{X_i, f_{ji}\}$, it suffices to verify the universal property of $\{X_\infty, f_{\infty i}\}$. Suppose Y is a compact Hausdorff space and $\{g_i: Y \to X_i\}$ is a family of continuous onto maps such that $f_{ji} g_j = g_i$ whenever $j \geq i$. Define $g: Y \to X_\infty$ by $(gy)_i = g_i y$. The fact that g is onto follows from the compactness of Y. Clearly $f_{\infty i} g = g_i$ for all $i \in D$. Suppose $f_{\infty i} h = g_i$ for all $i \in D$, where $h: Y \to X_\infty$. Then $h = g$. $\quad\square$

Proposition 26.2. *If $\{X_i, f_{ji}\}$ is an inverse system in $\mathcal{H}$, then* (i) *iff* (ii).

(i) x *is an isolated point of* $\varprojlim X_i$.

(ii) *There is an i such that $f_{\infty i} x$ is an isolated point of X_i and $f_{ji}^{-1} f_{\infty i} x$ has only one member for all $j \geq i$.*

Proof. Suppose $\{x\}$ is open. Thus there is an i and an open $U \subset X_i$ such that $\{x\} = f_{\infty i}^{-1}[U]$. Consequently $U = \{f_{\infty i} x\}$ and $f_{ji}^{-1} f_{\infty i} x = \{f_{\infty j} x\}$ for all $j \geq i$.

Suppose (ii). Then $\{f_{\infty i} x\}$ is open, $f_{\infty i}^{-1} f_{\infty i} x = \{x\}$, and $\{x\}$ is open. $\quad\square$

Proposition 26.3. *If $f: X \to Y$ belongs to $\mathcal{H}$ and is one-one, then f is a homeomorphism.*

Proof. Suppose $U \subset X$ is open. Let $\{V_i\}$ be an open cover of $Y - f[U]$. Then $\{f^{-1}[V_i]\}$ is an open cover of $X - U$. Since $X - U$ is compact, finitely many of the V_i's cover $Y - f[U]$. Thus $Y - f[U]$ is compact and $f[U]$ is open. $\quad\square$

Section 27

Towards Morley's Analysis of 1-Types

Let T be a substructure complete theory, and let $\mathcal{K}(T)$ be the category of all substructures of all models of T and all monomorphisms. Suppose $\mathcal{A} \subset \mathcal{B} \in \mathcal{K}(T)$. $\mathcal{B}$ is a simple extension of $\mathcal{A}$ if there is a $b \in B$ such that $\mathcal{B}$ is the least substructure of $\mathcal{B}$ whose universe contains $A \cup \{b\}$, in symbols $\mathcal{A}(b) = \mathcal{B}$. Two simple extensions of $\mathcal{A}$, say $\mathcal{A}(b)$ and $\mathcal{A}(c)$, are isomorphic over $\mathcal{A}$ if there exists an isomorphism

$$f \colon \mathcal{A}(b) \xrightarrow{\approx} \mathcal{A}(c)$$

such that $f \mid A = 1_A$ and $fb = c$. If $\mathcal{A}(b)$ and $\mathcal{A}(c)$ are isomorphic over $\mathcal{A}$, then b and c are said to be isomorphic over $\mathcal{A}$. There is at most one $f \colon \mathcal{A}(b) \xrightarrow{\approx} \mathcal{A}(c)$ such that $f \mid A = 1_A$ and $fb = c$, since each member of $\mathcal{A}(b)$ is named by some constant term $t(b)$, where $t(x)$ is a term in the language of $T \cup D\mathcal{A}$.

The Morley analysis of 1-types yields a classification of the isomorphism types of the simple extensions of $\mathcal{A}$. The classification extends the distinction between algebraic and transcendental; it will be of maximum interest when T is ω-stable. Define

$$S\mathcal{A} = S_1(T \cup D\mathcal{A}).$$

$T \cup D\mathcal{A}$ is a complete theory, since T is substructure complete. If $p \in S\mathcal{A}$, then there exists a simple extension $\mathcal{A}(b)$ such that b realizes p in every model of T extending $\mathcal{A}(b)$. If both b and c realize p, then there exists a unique isomorphism f: $\mathcal{A}(b) \xrightarrow{\approx} \mathcal{A}(c)$ such that $f \mid A = 1_A$ and $fb = c$. Conversely, if b and c are isomorphic over $\mathcal{A}$, then b and c realize the same l-type of $S\mathcal{A}$. Thus the substructure completeness of T allows equating the l-types of $S\mathcal{A}$ and the isomorphism types of the simple extensions of $\mathcal{A}$.

Each formula $F(x)$ in the language of $T \cup D\mathcal{A}$ gives rise to a subset of $S\mathcal{A}$ defined by

$$U_{F(x)} = \{p \mid F(x) \in p\}.$$

Let the collection of all such subsets of $S\mathcal{A}$ serve as a base for a topology of $S\mathcal{A}$. Then $S\mathcal{A}$ is a Stone space, i.e. a compact Hausdorff space whose clopen sets form a base for its topology. (A set is clopen if it is both open and closed.) The compactness of $S\mathcal{A}$ is an immediate consequence of 7.2.

Suppose $f: \mathcal{A} \to \mathcal{B}$ belongs to $\mathcal{K}(T)$. Then

$$Sf: S\mathcal{B} \to S\mathcal{A}$$

is defined by:

$$F(\underline{a}_1, \ldots, \underline{a}_n, x) \in (Sf)q \quad \text{iff} \quad F(\underline{fa}_1, \ldots, \underline{fa}_n, x) \in q.$$

Clearly Sf is a continuous onto map. Let

$$S: \mathcal{K}(T) \to \mathcal{H}$$

be the contravariant functor which assigns to each $\mathcal{A}$ of $\mathcal{K}(T)$, the compact Hausdorff space $S\mathcal{A}$ of $\mathcal{H}$, and to each $f: \mathcal{A} \to \mathcal{B}$ of $\mathcal{K}(T)$, the continuous onto map $Sf: S\mathcal{B} \to S\mathcal{A}$ of $\mathcal{H}$.

Proposition 27.1. (i) *If T is a substructure complete theory, then the category $\mathcal{K}(T)$ admits direct limits and filtrations with amalgamation, and the contravariant functor $S: \mathcal{K}(T) \to \mathcal{H}$ preserves limits.* (ii) *If T is complete, then $\mathcal{K}(T)$ admits filtrations.*

Proof. (i) $\mathcal{K}(T)$ admits filtrations with amalgamation by 13.1, and direct limits by 10.2. Let $\{\mathcal{A}_i, f_{ij}\}$ be a direct system of $\mathcal{K}(T)$. Then $\{S\mathcal{A}_i, Sf_{ij}\}$ is an inverse system of $\mathcal{H}$. $\varprojlim S\mathcal{A}_i$ exists by 26.1. Let $g_i \colon \varprojlim S\mathcal{A}_i \to S\mathcal{A}_i$ be such that $(Sf_{ij})g_j = g_i$ when $i \leq j$. Let $\mathcal{A}_\infty = \varinjlim \mathcal{A}_i$ and $f_{i\infty} \colon \mathcal{A}_i \to \mathcal{A}_\infty$. There exists a unique

$$h \colon S\mathcal{A}_\infty \to \varprojlim S\mathcal{A}_i$$

such that $g_i h = Sf_{i\infty}$ for all i. If h is one-one, then h is a homeomorphism by 26.3. Suppose $hp = hq$. Then $g_i hp = g_i hq$, and so $Sf_{i\infty}p = Sf_{i\infty}q$ for all i. But then $p = q$.

(ii) $\mathcal{K}(T)$ admits filtrations by 7.7. $\qquad\qquad\square$

Section 28

The Cantor–Bendixson Derivative

The Morley derivative is an improved version of the Cantor–Bendixson derivative, improved in a sense to be made precise at the beginning of Sec. 29.

Suppose $X \in \mathcal{H}$; x is an isolated point of X if $\{x\}$ is an open subset of X; x is a limit point if x is not isolated. The Cantor–Bendixson derivative of X, denoted by dX, is the set of all limit points of X. Clearly $dX \in \mathcal{H}$. For each ordinal α, the α-th Cantor–Bendixson derivative of X, denoted by $d^\alpha X$, is defined by induction:

$$d^0 X = X$$
$$d^{\alpha+1} = d(d^\alpha X)$$
$$d^\lambda X = \cap \{d^\alpha X \mid \alpha < \lambda\}.$$

For every α, $d^\alpha X \in \mathcal{H}$. The Cantor–Bendixson rank of X, denoted by α_X, is the least α such that $d^\alpha X$ has no isolated points. If $x \in X$ and for some α, necessarily unique, $x \in d^\alpha X - d^{\alpha+1} X$ then x is said to have Cantor–Bendixson rank α.

Theorem 28.1 (Cantor, Bendixson). *Suppose* $X \in \mathcal{H}$. *(1) If X has a countable base, then $\alpha_X < \omega_1$. (2) If X is*

*countable, then every point of X has a Cantor–Bendixson rank.
(3) If every point of X has a Cantor–Bendixson rank, then the
isolated points of X are dense in X.*

Proof.   (1) Suppose X has a countable base. For each $\alpha < \alpha_X$,
choose an $x_\alpha \in X$ and a basic open U_α such that $d^\alpha X \cap U_\alpha = \{x_\alpha\}$. Then $\alpha < \beta < \alpha_X$ implies $x_\beta \notin U_\alpha$ and $U_\alpha \neq U_\beta$. So
card α_X is countable.

(2) Suppose the perfect kernel of X, call it X_0^0, is nonempty.
Thus X_0^0 is a nonempty closed subset of X without isolated
points. Fix $i \geq 0$ and $j < 2^i$. Suppose X_j^i has the properties
claimed for X_0^0. Let $x, y \in X_j^i$ be distinct. Since $X \in \mathcal{H}$, X
is normal; i.e. disjoint closed subsets of X can be separated by
disjoint open subsets of X. It follows there exist disjoint closed
subsets of X_j^i, call them X_{2j}^{i+1} and X_{2j+1}^{i+1}, such that $x \in X_{2j}^{i+1}$,
$y \in X_{2j+1}^{i+1}$, and neither X_{2j}^{i+1} nor X_{2j+1}^{i+1} have isolated points.

Let $t \colon \omega \to \omega$ be a function such that $t(0) = 0$ and $t(i+1) \in \{2t(i), 2t(i)+1\}$. For each such t,

$$\cap \{X_{t(i)}^i \mid i < \omega\}$$

is nonempty. Distinct t's give rise to distinct intersections. But
then X is uncountable.

(3) Suppose every point of X has a Cantor–Bendixson rank.
Let U be a nonempty open subset of X. Choose $x \in U$ so that
no member of U has lower rank than x. There exists an open V
such that $\{x\} = V \cap d^\alpha X$, where α is the rank of x. But then
$\{x\} = U \cap V$. $\square$

Proposition 28.2. *Let $f \colon Y \to X$ belong to $\mathcal{H}$. Then
$f[dY] \supset dX$.*

Proof.   Fix $x \in X$. Suppose $f^{-1}(x)$ contains no limit points
of Y. Then $f^{-1}(x)$ is open, $Y - f^{-1}(x)$ is compact, $X - \{x\}$ is
compact, and $\{x\}$ is open. $\square$

Exercise 28.3. Suppose $\mathcal{A} \in \mathcal{K}(T)$ and $\mathcal{A}(b) = \mathcal{A}(c)$. Let $q \in S\mathcal{A}$ (respectively $r \in S\mathcal{A}$) be the l-type realized by b (respectively c). Show q and r have the same Cantor–Bendixson rank if either one has a Cantor–Bendixson rank.

Section 29

The Morley Derivative

The Cantor–Bendixson derivative d is imperfect in the sense that the conclusion of 28.2 cannot be improved to read $f[dY] = dX$. The Morley derivative D is designed so that $f[DY] = DX$ whenever $f \colon Y \to X$ belongs to $\mathcal{H}$.

Let $F \colon \mathcal{K} \to \mathcal{H}$ be a contravariant functor with the properties attributed by 27.1 to $S \colon \mathcal{K}(T) \to \mathcal{H}$ when T is a substructure complete theory. Thus $\mathcal{K}$ admits direct limits and filtrations with amalgamation, and F preserves limits. Let $\mathcal{A} \in \mathcal{K}$. Then $x \in DF\mathcal{A}$ iff $x \in F\mathcal{A}$ and some $f \colon \mathcal{A} \to \mathcal{B}$ belonging to $\mathcal{K}$ has the property that $(Ff)^{-1}x$ contains a limit point. For every $f \colon \mathcal{A} \to \mathcal{B}$ belonging to $\mathcal{K}$, define DFf to be the restriction of Ff to $DF\mathcal{B}$.

Proposition 29.1. $DF \colon \mathcal{K} \to \mathcal{H}$ *is a contravariant functor that preserves limits.*

Proof. Suppose $f \colon \mathcal{A} \to \mathcal{B}$ belongs to $\mathcal{K}$. To see that DF is a contravariant functor, it suffices to see that DFf maps $DF\mathcal{B}$ onto $DF\mathcal{A}$. Fix $y \in DF\mathcal{B}$. For some $g \colon \mathcal{B} \to \mathcal{C}$, $(Fg)^{-1}y$ contains a limit point. But then $(Fgf)^{-1}Ffy$ contains a limit point, and so $DFfy \in DF\mathcal{A}$.

Fix $x \in DF\mathcal{A}$ in the hope of finding a $y \in DF\mathcal{B}$ such that $DFfy = x$. Choose $g \colon \mathcal{A} \to \mathcal{C}$ so that $(Fg)^{-1}x$ contains a limit point. Since $\mathcal{K}$ admits filtrations with amalgamation, the following diagram can be completed as shown.

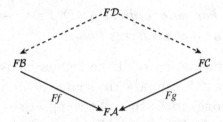

By 28.2 x has a pre-image z that is a limit point of $F\mathcal{D}$. The image of z in $F\mathcal{B}$, call it y, belongs to $DF\mathcal{B}$, and $DFfy = x$.

Suppose $\{\mathcal{A}_i, f_i\}$ is a direct system in $\mathcal{K}$ whose direct limit consists of $\mathcal{A}_\infty$ and $\{f_i: \mathcal{A}_i \to \mathcal{A}_\infty\}$. Let the inverse limit of $F\mathcal{A}_i, Ff_{ij}$ consist of $\varprojlim F\mathcal{A}_i$ and $\{g: \varprojlim F\mathcal{A}_i \to F\mathcal{A}_i\}$. Since F preserves limits, the unique

$$g: F\mathcal{A}_\infty \to \varprojlim F\mathcal{A}_i$$

such that $g_i g = Ff_i$ for all i is a homeomorphism. Let the inverse limit of $\{DF\mathcal{A}_i, DFf_{ij}\}$ consist of $\varprojlim DF\mathcal{A}_i$ and $\{h_i : \varprojlim DF\mathcal{A}_i \to DF\mathcal{A}_i\}$. Let

$$h: DF\mathcal{A}_\infty \to \varprojlim DF\mathcal{A}_i$$

be the unique map such that $h_i h = DFf_i$ for all i. If h is one-one, then h is a homeomorphism by 26.3. Suppose $hx = hy$. Then for all i, $DFf_i x = DFf_i y$, $Ff_i x = Ff_i y$ and $g_i g x = g_i g y$. Consequently $gx = gy$. Then $x = y$ since g is a homeomorphism. $\qquad\square$

The α-th Morley derivative of F, denoted $D^\alpha F$, is defined by induction:

$$D^0 F = F.$$
$$D^{\alpha+1} F = D(D^\alpha F).$$
$$D^\lambda F\mathcal{A} = \cap \{D^\alpha F\mathcal{A} \mid \alpha < \lambda\}.$$

If $f: \mathcal{A} \to \mathcal{B}$ belongs to $\mathcal{K}$, then $D^\lambda Ff$ is the restriction of Ff to $D^\lambda F\mathcal{B}$.

Lemma 29.2. *For each ordinal* α, $D^\alpha F: \mathcal{K} \to \mathcal{H}$ *is a contravariant functor that preserves limits.*

Proof. By induction on α. If the lemma holds for α, then it holds for $\alpha+1$ by 29.1. Suppose the lemma holds for every $\alpha < \lambda$. If $f: \mathcal{A} \to \mathcal{B}$ belongs to $\mathcal{K}$, then the compactness of $F\mathcal{B}$ implies that $D^\lambda Ff$ maps $D^\lambda F\mathcal{B}$ onto $D^\lambda F\mathcal{A}$.

Suppose $\{\mathcal{A}_i, f_{ij}\}$ is a direct system in $\mathcal{K}$ whose direct limit consists of $\mathcal{A}_\infty$ and $\{f_i: \mathcal{A}_i \to \mathcal{A}_\infty\}$. Let the inverse limit of $\{D^\lambda F\mathcal{A}_i, D^\lambda Ff_{ij}\}$ consist of $\varprojlim D^\lambda F\mathcal{A}_i$ and $\{h_i: \varprojlim D^\lambda F\mathcal{A}_i \to D^\lambda F\mathcal{A}_i\}$. There exists a unique

$$h: D^\lambda F\mathcal{A}_\infty \to \varprojlim D^\lambda F\mathcal{A}_i$$

such that $h_i h = D^\lambda Ff_i$ for all i. If h is one-one, then h is a homeomorphism by 26.3. Suppose $hx = hy$. Then $x, y \in F\mathcal{A}_\infty$ and $Ff_i x = Ff_i y$ for all i. Let the inverse limit of $\{F\mathcal{A}_i, Ff_{ij}\}$ consist of $\varprojlim F\mathcal{A}_i$ and $\{g_i: \varprojlim F\mathcal{A}_i \to F\mathcal{A}_i\}$. The unique

$$g: F\mathcal{A}_\infty \to \varprojlim F\mathcal{A}_i$$

such that $g_i g = Ff_i$ for all i is a homeomorphism. So $g_i gx = g_i gy$ for all i, $gx = gy$ and $x = y$. $\qquad\square$

Suppose there is an α (necessarily unique) such that $x \in D^\alpha F\mathcal{A} - D^{\alpha+1}F\mathcal{A}$. Then x is said to be a (Morley) ranked point of $F\mathcal{A}$ of rank α.

Proposition 29.3 (Rank Rule). *Let* $f: \mathcal{A} \to \mathcal{B}$ *belong to* $\mathcal{K}$, *and let* x *be a ranked point of* $F\mathcal{A}$. *Suppose* $Ffy = x$. *Then* y *is a ranked point of* $F\mathcal{B}$ *and rank* $y \leq$ *rank* x.

Proof. Let rank $x = \alpha$. Thus $x \in D^\alpha F\mathcal{A} - D^{\alpha+1}F\mathcal{A}$. By 29.2 Ff maps $D^{\alpha+1}F\mathcal{B}$ onto $D^{\alpha+1}F\mathcal{A}$, so $y \notin D^{\alpha+1}F\mathcal{B}$. $\qquad\square$

Proposition 29.4. *Let $x \in FA - DFA$. Then there exists a positive integer n such that for all $f: A \to B$ in K, the cardinality of $(Ff)^{-1}x$ is at most n.*

Proof. Suppose not. A direct system $\{A_n \mid n < \omega\}$ is defined by induction on n. $A_0 = A$. Suppose

$$A_0 \xrightarrow{f_0} A_1 \xrightarrow{f_1} \cdots \xrightarrow{f_{n-1}} A_n$$

has already been defined. Choose $f: A_0 \to B$ so that $(Ff)^{-1}x$ has cardinality at least n. Since K admits filtrations with amalgamation, the above diagram can be completed as shown.

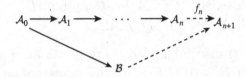

Clearly $(Ff_n \cdots f_1 f_0)^{-1}x$ has cardinality at least n. Let $A_\infty = \varinjlim A_i$. Then $(Ff_{0\infty})^{-1}x$ is infinite and so must contain a limit point. But then $x \in DFA$. $\qquad\square$

Suppose x is a ranked point of FA of rank α. Proposition 29.4 applies to every $F: K \to H$ that preserves limits, and in particular to $D^\alpha F$ by 29.2. Consequently there is an n such that for all $f: A \to B$, the cardinality of $(D^\alpha Ff)^{-1}x$ is at most n. The least such n, denoted by $\deg x$, is called the degree of x.

Proposition 29.5 (Degree Rule). *If x is a ranked point of FA and $f: A \to B$, then*

$$\deg x = \sum \{\deg y \mid Ffy = x \ \& \ \text{rank } y = \text{rank } x\}.$$

Proof. By 29.2 it is safe to assume rank $x = 0$. Choose $g: A \to C$ so that the cardinality of $(Fg)^{-1}x$ is $\deg x$. Since K admits filtrations with amalgamation, the following diagram can be completed as shown.

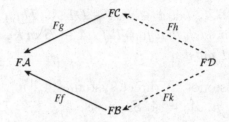

Then $\deg x = \ \text{card} \ (Fhg)^{-1}x = \sum\{\deg y \mid Ffy = x\}$. $\square$

Proposition 29.6. *Suppose α is an ordinal such that $D^\alpha F\mathcal{A} = 0$ for some $\mathcal{A} \in \mathcal{K}$. If $\mathcal{K}$ admits filtrations, then $D^\alpha F\mathcal{B} = 0$ for all $\mathcal{B} \in \mathcal{K}$.*

Proof.   There exist $\mathcal{C}$, $f\colon \mathcal{A} \to \mathcal{C}$ and $g\colon \mathcal{B} \to \mathcal{C}$. By 29.2 $D^\alpha F\mathcal{B} = (D^\alpha Fg)(D^\alpha Ff)^{-1}D^\alpha F\mathcal{A}$. $\square$

F is totally transcendental if there exists an α such that for all $\mathcal{B} \in \mathcal{K}$, $D^\alpha F\mathcal{B} = 0$. If F is totally transcendental, then the rank and degree rules are helpful in the study of $x \in F\mathcal{A}$. The rank rule says no pre-image of x can have higher rank than x. Suppose $f\colon \mathcal{A} \to \mathcal{B}$ belongs to $\mathcal{K}$. The degree rule implies: if $\deg x = 1$, then x has a unique pre-image in $F\mathcal{B}$ of the same rank and degree as x. This last fact is central to the proof of Theorem 35.6.

Exercise 29.7. Suppose T is a complete, substructure complete theory. Let $S\colon \mathcal{K}(T) \to \mathcal{H}$ be the contravariant functor mentioned in 27.1. Assume S is totally transcendental. Show there exists a countable ordinal α such that for all $\mathcal{A} \in \mathcal{K}(T)$, $D^\alpha S\mathcal{A} = 0$.

Exercise 29.8. Let $S\colon \mathcal{K}(\mathrm{ACF}_0) \to \mathcal{H}$. Show S is totally transcendental.

Section 30

Autonomous Subcategories

Throughout this section $\mathcal{K}$ is a category that admits direct limits and filtrations with amalgamation, and $F\colon \mathcal{K} \to \mathcal{H}$ is a contravariant functor that preserves limits. Fix $\mathcal{A} \in \mathcal{K}$. The computation of $D^\alpha F\mathcal{A}$, the α-th Morley derivative of $F\mathcal{A}$, is in practice cumbersome, because its definition ranges over all of $\mathcal{K}$. In this section and the next conditions are given under which the value of $D^\alpha F\mathcal{A}$ depends only on $\mathcal{K}^*$, a small full subcategory of $\mathcal{K}$ containing $\mathcal{A}$. The reduction of $\mathcal{K}$ to $\mathcal{K}^*$ is inspired by the downward going Skolem–Löwenheim theorem. The reduction will be at its most extreme in Sec. 31, where it will be seen that for many $\mathcal{A}$'s, the α-th Morley derivative of $S\mathcal{A}$ equals the α-th Cantor–Bendixson derivative of $S\mathcal{A}$ for all α.

From now on $\mathcal{K}^*$ is a full subcategory of $\mathcal{K}$. $F^*\colon \mathcal{K}^* \to \mathcal{H}$ is the restriction of $F\colon \mathcal{K} \to \mathcal{H}$ to $\mathcal{K}^*$. $\mathcal{K}^*$ is autonomous if $(DF)^* = D(F^*)$ for every $F\colon \mathcal{K} \to \mathcal{H}$.

Proposition 30.1. *If $\mathcal{K}^*$ is autonomous, then $(D^\alpha F)^* = D^\alpha(F^*)$ for all α.*

Proof. By induction on α. Let $\mathcal{A} \in \mathcal{K}^*$. Suppose $(D^\alpha F)^* = D^\alpha(F^*)$. Then $(D^{\alpha+1}F)^*\mathcal{A} = D((D^\alpha F)\mathcal{A}) = D((D^\alpha F)^*\mathcal{A}) = D^{\alpha+1}(F^*)\mathcal{A}$. Suppose $(D^\alpha F)^* = D^\alpha(F^*)$ for all $\alpha < \lambda$. Then

117

$(D^\lambda F)^* \mathcal{A} = D^\lambda F \mathcal{A} = \cap \{D^\alpha F \mathcal{A} \,|\, \alpha < \lambda\} = \cap \{D^\alpha (F^*) \mathcal{A} \,|\, \alpha < \lambda\} = D^\lambda (F^*) \mathcal{A}.$ $\qquad\qquad\qquad\qquad\qquad\qquad$ □

Let $\{X_i \,|\, i \in D\}$ be an inverse system of $\mathcal{H}$ directed by $\langle D, \leq \rangle$. Suppose $\langle E, \leq \rangle \subset \langle D, \leq \rangle$ and $\{X_i \,|\, i \subset E\}$ is an inverse system directed by $\langle E, \leq \rangle$. Let the inverse limit of $\{X_i \,|\, i \in E\}$ consist of $\varprojlim \{X_i \,|\, i \in E\}$ and maps

$$f_{\infty i}^F : \varprojlim \{X_i \,|\, i \in E\} \to X_i.$$

There exists a unique map

$$f^{DE} : \varprojlim \{X_i \,|\, i \in D\} \to \varprojlim \{X_i \,|\, i \in E\}$$

with the property that $f_{\infty i}^E f^{DE} = f_{\infty i}^D$ for all $i \in E$.

Proposition 30.2. *Let $\{X_i \,|\, i \in D\}$ be an inverse system of $\mathcal{H}$ directed by $\langle D, \leq \rangle$. Suppose x is a limit point of $\varprojlim \{X_i \,|\, i \in D\}$. Then there exists a countable $E \subset D$ such that $f^{DE} x$ is a limit point of $\varprojlim \{X_i \,|\, i \in E\}$.*

Proof. A sequence $i_0 \leq i_1 \leq i_2 \leq \cdots$ of members of D is defined by induction. i_0 is chosen artlessly. Fix $k \geq 0$. If $f_{\infty i_k}^D x$ is a limit point of X_{i_k}, choose $i_{k+1} \geq i_k$. If $f_{\infty i_k}^D x$ is an isolated point of X_{i_k}, choose $i_{k+1} \geq i_k$ so that

$$f_{i_{k+1} i_k}^{-1} (f_{\infty i_k}^D x)$$

has at least two members in $X_{i_{k+1}}$; if no such choice could be made, then x would be isolated. Let $E = \{i_k \,|\, k < \omega\}$. Suppose $f^{DE} x$ were isolated. Then by 26.2 there would be a k such that $f_{\infty i_k}^D x$ is isolated and $f_{i_{k+1} i_k}^{-1} (f_{\infty i_k}^D x)$ has only one member. $\qquad$ □

The above proposition is reminiscent of the downward going Skolem–Löwenheim theorem.

A direct system $\{\mathcal{A}_i \,|\, i \in D\}$ of $\mathcal{K}$ is countable if D is countable. $\mathcal{K}^*$ is closed under formation of limits in $\mathcal{K}$ of countable direct systems, if every countable direct system in $\mathcal{K}^*$ has a direct limit in $\mathcal{K}^*$ isomorphic to its direct limit in $\mathcal{K}$. $\mathcal{K}^*$ is dense in $\mathcal{K}$ if for every $f: \mathcal{A} \to \mathcal{B}$ such that $\mathcal{A} \in \mathcal{K}^*$ and $\mathcal{B} \in \mathcal{K}$, there is a direct system $\{\mathcal{B}_i\}$ in $\mathcal{K}^*$ such that $\mathcal{B}(=\mathcal{B}_\infty)$ is the direct limit in $\mathcal{K}$ of $\{\mathcal{B}_i\}$, and such that for some j, $\mathcal{A} = \mathcal{B}_j$ and $f = f_{j\infty}$.

Theorem 30.3 (S. Simpson). *Suppose $\mathcal{K}^*$ is closed under formation of limits in $\mathcal{K}$ of countable direct systems. If $\mathcal{K}^*$ is dense in $\mathcal{K}$, then $\mathcal{K}^*$ is autonomous.*

Proof. Suppose $F: \mathcal{K} \to \mathcal{H}$ and $\mathcal{A} \in \mathcal{K}^*$. Clearly $D(F^*)\mathcal{A} \subset (DF)^*\mathcal{A}$. Let $y \in (DF)^*\mathcal{A} = DF\mathcal{A}$. Then for some $f: \mathcal{A} \to \mathcal{B}$ in $\mathcal{K}$, there is a limit point x of $DF\mathcal{B}$ such that $Ffx = y$. Choose a direct system $\{\mathcal{B}_i \,|\, i \in D\}$ of $\mathcal{K}^*$ such that $\mathcal{B}$ is the direct limit in $\mathcal{K}$ of $\{\mathcal{B}_i\}$, and such that for some $j \in D$, $\mathcal{A} = \mathcal{B}_j$ and $f = f_{j\infty}$. By 30.2 there is a countable $E \subset D$ and a limit point

$$z \in \varprojlim\{F\mathcal{B}_i \,|\, i \in E\}$$

such that the map

$$f_{\infty j}^E: \varprojlim\{F\mathcal{B}_i \,|\, i \in E\} \to F\mathcal{B}_j$$

takes z to y. But $\varprojlim\{F\mathcal{B}_i \,|\, i \in E\}$ belongs to $\mathcal{K}^*$, so $y \in D(F^*)\mathcal{A}$. $\qquad\square$

Suppose T is a substructure complete theory. For each uncountable cardinal κ, let $\mathcal{K}_\kappa(T)$ be the category of all substructures of all models of T of cardinality less than κ and all associated monomorphisms.

Corollary 30.4. *If $\kappa > \omega$, then $\mathcal{K}_\kappa(T)$ is an autonomous subcategory of $\mathcal{K}(T)$.*

Proof. It is safe to assume κ is regular, since the union of autonomous subcategories is autonomous. It is immediate from the uncountability and regularity of κ that $\mathcal{K}_\kappa(T)$ is closed under formation of limits in $\mathcal{K}(T)$ of countable direct systems. To see that $\mathcal{K}_\kappa(T)$ is dense in $\mathcal{K}(T)$, fix $\mathcal{A} \subset \mathcal{B}$ such that $\mathcal{A} \in \mathcal{K}_\kappa(T)$ and $\mathcal{B} \in \mathcal{K}(T)$, and let $\{\mathcal{B}_i\}$ be the set of all $\mathcal{D} \in \mathcal{K}_\kappa(T)$ such that $\mathcal{A} \subset \mathcal{D} \subset \mathcal{B}$. Direct $\{\mathcal{B}_i\}$ by set inclusion. Then $\mathcal{B} = \varinjlim\{\mathcal{B}_i\}$.

$\square$

It can be shown for every countable T that $\mathcal{K}(T)$ has a countable autonomous subcategory. All the ideas needed to prove this last fact can be found in the next section.

Section 31

Bounds on the Ranks of
1-Types

Throughout this section T is a complete, substructure complete theory. Let $S\colon \mathcal{K}(T) \to \mathcal{H}$ be the contravariant Stone functor defined in Sec. 27. Thus for each $\mathcal{A}$ a substructure of a model of T, $S\mathcal{A}$ is the Stone space whose points correspond to the isomorphism types over $\mathcal{A}$ of the simple extensions of $\mathcal{A}$. For each ordinal α, define $D^\alpha S$, the α-th Morley derivative of S, as in Sec. 29. If there exists an α such that $p \in D^\alpha S\mathcal{A} - D^{\alpha+1}S\mathcal{A}$, then p is a ranked point of $S\mathcal{A}$ of rank α. In symbols: rank $p = \alpha$. If p is ranked, then p has a degree, denoted by deg p, as defined in Sec. 29. A downward going Skolem–Löwenheim argument (31.5), somewhat obscured, shows rank $p < \omega_1$.

Suppose $\mathcal{U}, \mathcal{V}, \mathcal{W} \in \mathcal{K}(T)$. $\mathcal{V}$ is finitely generated if there exists a finite $Y \subset V$ such that $\mathcal{V}$ is the least substructure of $\mathcal{V}$ whose universe contains Y. Suppose $x \in D^\alpha S\mathcal{V}$ and $i\colon \mathcal{V} \subset \mathcal{W}$ is the inclusion map; x splits in $D^\alpha S\mathcal{W}$ if $(D^\alpha Si)^{-1}x$ has at least two members.

$\mathcal{U}$ is a β-universal domain for T if for every $\alpha < \beta$, every finitely generated $\mathcal{V} \subset \mathcal{U}$, and every isolated $x \in D^\alpha S\mathcal{V}$, the following holds: if x splits in $D^\alpha S\mathcal{W}$ for some $\mathcal{W} \in \mathcal{K}(T)$, then x splits in $D^\alpha S\mathcal{W}^*$ for some finitely generated $\mathcal{W}^* \subset \mathcal{U}$. $\mathcal{U}$ is a universal domain for T if $\mathcal{U}$ is $\alpha_{\mathcal{U}}$-universal. ($\alpha_{\mathcal{U}}$ is the Cantor–Bendixson rank of $\mathcal{U}$.)

If $\mathcal{U}$ is ω-saturated, then $\mathcal{U}$ is a universal domain, but not conversely. T may not have a countable ω-saturated model, but T must have a countable universal domain according to 31.2.

Suppose $\mathcal{V} \subset \mathcal{W}^* \in \mathcal{K}(T)$; $\mathcal{W}^*$ is finitely generated over $\mathcal{V}$ if there exists a finite $Y \subset W^*$ such that $\mathcal{W}^*$ is the least substructure of $\mathcal{W}^*$ whose universe contains $V \cup Y$. In symbols: $\mathcal{W}^* = \mathcal{V}(Y)$ or $\mathcal{W}^* = \mathcal{V}(y_1, \ldots, y_n)$, where $Y = \{y_1, \ldots, y_n\}$.

Proposition 31.1. *If $x \in DS^\alpha \mathcal{V}$ and x splits in $D^\alpha S\mathcal{W}$ for some $\mathcal{W} \in \mathcal{K}(T)$, then x splits in $D^\alpha S\mathcal{W}^*$ for some $\mathcal{W}^*$ finitely generated over $\mathcal{V}$.*

Proof. Let $\{\mathcal{V}_i\}$ be the set of all $\mathcal{V}_i \subset \mathcal{W}$ such that $\mathcal{V}_i$ is finitely generated over $\mathcal{V}$. Direct $\{\mathcal{V}_i\}$ by set inclusion. Then $\mathcal{W} = \lim_{\rightarrow} \mathcal{V}_i$. By 29.2 $D^\alpha S\mathcal{W} = \lim_{\leftarrow} D^\alpha S\mathcal{V}_i$. But then 26.2 implies x splits in $D^\alpha S\mathcal{V}_i$ for some i. $\qquad\square$

Proposition 31.2. *If $\mathcal{A} \in \mathcal{K}(T)$ is countable, then there exists a countable universal domain $\mathcal{U} \supset \mathcal{A}$.*

Proof. Fix $\beta < w_1$. A countable, β-universal domain can be constructed as follows. $\mathcal{U}_o = \mathcal{A}$. $\mathcal{U} = \cup\{\mathcal{U}_n | n < \omega\}$. For each $\alpha < \beta$, each finitely generated $\mathcal{V} \subset \mathcal{U}_n$, and each isolated $x \in D^\alpha S\mathcal{V}$, use 31.1 to choose a $\mathcal{W}^\mathcal{V}_{\alpha,x}$ finitely generated over $\mathcal{V}$ such that: if x splits in $D^\alpha S\mathcal{W}$ for some $\mathcal{W} \in \mathcal{K}(T)$, then x splits in $\mathcal{W}^\mathcal{V}_{\alpha,x}$. Let $\mathcal{U}_{n+1}$ be a countable member of $\mathcal{K}(T)$ whose universe contains

$$U_n \cup \{\mathrm{W}^\mathcal{V}_{\alpha,x} | \alpha, \mathcal{V}, x\}.$$

β-unversality leads to universality as follows. There exists a countable

$$L^1 = L(\alpha_1, T, \mathcal{A})$$

such that L^1 is a Σ_1 substructure of M, the class of all sets; in addition L^1 satisfies Σ_2 replacement and every member of L^1 is countable in L^1.

Suppose $\mathcal{B} \in \mathcal{K}(T) \cap L^1$. The definition of

$$(S\mathcal{B} - D^\beta S\mathcal{B})$$

is a Π_1^1, hence Σ_1, recursion in M that executes correctly in L^1 for all $\beta < \alpha_1$. As in 30.4 only countable superstructures of $\mathcal{B}$ matter, and non-isolated pre-images exist in M iff they exist in L^1. It follows that for each $\beta < \alpha_1$, there exists a β-universal domain for T in L^1. By Barwise compactness there exists a countable, α_1-universal domain $\mathcal{U}$. By effective type omitting $\mathcal{U}$ can be defined so that $\omega_1^{\mathcal{U}} \leq \alpha_1$. ($\omega_1^{\mathcal{U}}$ is the least α such that $L(\alpha, \mathcal{U})$ satisfies Σ_1 replacement.) But then a result of Kreisel implies $\alpha_{\mathcal{U}}$, the Cantor–Bendixson rank of $\mathcal{U}$, is at most $\omega_1^{\mathcal{U}}$, and so $\mathcal{U}$ is a universal domain. $\qquad\square$

Lemma 31.3. *Let $\mathcal{U}$ be a universal domain for T. Then the Morley analysis of $S\mathcal{U}$ coincides with its Cantor–Bendixson analysis; i.e. for each α,*

$$D^\alpha S\mathcal{U} = d^\alpha S\mathcal{U}.$$

Proof. By induction on α. Suppose $D^\alpha S\mathcal{U} = d^\alpha S\mathcal{U}$. Clearly $D^{\alpha+1} S\mathcal{U} \supset d^{\alpha+1} S\mathcal{U}$. Suppose $x \in D^{\alpha+1} S\mathcal{U}$ in the hope of showing $x \in d^{\alpha+1} S\mathcal{U}$. Clearly $x \in D^\alpha S\mathcal{U}$. Assume x is an isolated point of $D^\alpha S\mathcal{U}$, since otherwise $x \in d^{\alpha+1} S\mathcal{U}$. Then $\alpha < \alpha_{\mathcal{U}}$. It follows there is a $\mathcal{W} \in \mathcal{K}(T)$ such that x splits in $D^\alpha S\mathcal{W}$. Let $\{\mathcal{V}_i\}$ be the direct system of all finitely generated substructures of $\mathcal{U}$ directed by inclusion. Then $D^\alpha S\mathcal{U} = \varprojlim\{D^\alpha S\mathcal{V}_i\}$ by 29.2. Use 26.2 to choose $f_i: \mathcal{V}_i \subset \mathcal{U}$ so that $D^\alpha S f_i x$ is an isolated point of $D^\alpha S\mathcal{V}_i$ and $D^\alpha S f_i x$ does not split in $D^\alpha S\mathcal{U}$. Since $D^\alpha S f_i x$ splits

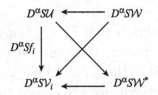

in $D^\alpha SW$, and since $\mathcal{U}$ is a universal domain, there must be a finitely generated $W^* \subset \mathcal{U}$ such that $D^\alpha S f_i x$ splits in $D^\alpha SW^*$. But then $D^\alpha S f_i x$ splits in $D^\alpha S\mathcal{U}$. $\square$

The Morley rank of T, denoted by α_T, is the least α such that $D^\alpha S\mathcal{A} = D^{\alpha+1} S\mathcal{A}$ for all finitely generated $\mathcal{A} \in \mathcal{K}(T)$.

Proposition 31.4. $D^{\alpha_T} S\mathcal{A} = D^{\alpha_T+1} S\mathcal{A}$ *for every* $\mathcal{A} \in \mathcal{K}(T)$.

Proof. Let $\{\mathcal{A}_i\}$ be the direct system of all finitely generated substructures of $\mathcal{A}$ directed by inclusion. Then $D^\alpha S\mathcal{A} = \lim_{\leftarrow} D^\alpha S\mathcal{A}_i$ by 29.2. $\square$

Theorem 31.5 (A. H. Lachlan). (i) *If* $p \in S\mathcal{A}$ *is ranked, then* rank $p < \omega_1$. (ii) $\alpha_T \leq \omega_1$.

Proof. (ii) is an immediate consequence of (i). To prove (i) construe $\mathcal{A}$ as the direct limit of its finitely generated substructures. By 29.2 there exists a $g \colon \mathcal{B} \subset \mathcal{A}$ such that $\mathcal{B}$ is finitely generated and rank $(Sg)p =$ rank p. By 31.2 there exists an $h \colon \mathcal{B} \subset \mathcal{U}$ such that $\mathcal{U}$ is a countable universal domain. The degree rule (29.5) supplies a $q \in S\mathcal{U}$ such that $Shq = Sgp$ and rank $q =$ rank Sgp. It follows from 31.3 that the rank of q equals its Cantor–Bendixson rank; the latter is countable by 28.1(1). $\square$

There is an effective version of 31.5(i) whose proof [Sa1] is a careful delineation of the absolute nature of Morley rank: if $p \in S\mathcal{A}$ has a rank and $\mathcal{A}$ is countable, then the Morley rank of p is an ordinal recursive in $\mathcal{A}$. (Since $\mathcal{A}$ is countable, its diagram can be encoded by some real number $\mathcal{A}^e$. An ordinal is recursive in $\mathcal{A}$ if it is isomorphic to some wellordering of ω recursive in $\mathcal{A}^e$.) If $0 < \alpha \leq \omega_1$, then there exists a T such that $\alpha = \alpha_T$. J. Baldwin [Bal] has shown that $\alpha_T < \omega$ for every T categorical in some uncountable power. Baldwin's striking result appears plausible when viewed in the light of Lemma 39.8. If $0 < n < \omega$, then there exists an ω_1-categorical T such that $\alpha_T = n$.

T is a totally transcendental theory if the functor $S: \mathcal{K}(T) \to \mathcal{H}$ is totally transcendental as defined in Sec. 29.

Theorem 31.6 (M. Morley). *T is totally transcendental iff T is ω-stable.*

Proof. Suppose T is totally transcendental and $\mathcal{A} \in \mathcal{K}(T)$ is countable. For each $p \in S\mathcal{A}$, choose a basic open $N_p \subset S\mathcal{A}$ such that $\{p\} = D^\alpha S\mathcal{A} \cap N_p$ for some α. If $p \neq q$, then $N_p \neq N_q$. It follows that $S\mathcal{A}$ is countable, since $S\mathcal{A}$ has a countable base.

Suppose T is ω-stable. By 19.6 there exists a countable ω-saturated model of T. By 28.1(2) every member of $S\mathcal{U}$ has a Cantor–Bendixson rank, hence a (Morley) rank by 31.3. It follows from 29.6 and 27.1(ii) that T is totally transcendental. $\square$

The importance of Theorem 31.6, and other theorems of the same type [Sh2], cannot be overestimated. Suppose T is ω-stable. Then 31.6 says every simple extension of every substructure of every model of T can be assigned a rank. The rank and degree rules of Sec. 29 make it possible when rank is omnipresent, to prove theorems about models of T by induction on rank, e.g. the proof of Shelah's uniqueness theorem (36.2). Even when T is not ω-stable the ranked types of T are often ubiquitous enough to make induction on rank a useful method for studying T. T is quasi-totally transcendental (q.t.t.) if for every $\mathcal{A} \in \mathcal{K}(T)$, the ranked points of $S\mathcal{A}$ are dense in $S\mathcal{A}$. Many theorems (e.g. 36.2) initially proved for totally transcendental theories have turned out to hold for q.t.t. theories as well. The notion of quasi-total transcendentality is absolute.

Suppose T is q.t.t. The density number (L. Blum) of T, denoted by d_T, is the least ordinal γ with the following property: for every $\mathcal{A} \in \mathcal{K}(T)$ the points of $S\mathcal{A}$ of Morley rank $< \gamma$ are dense in $S\mathcal{A}$. There exists a q.t.t. theory T such that $\alpha_T = \omega_1$; nonetheless d_T is countable for every q.t.t. T. (In fact d_T is recursive in T.)

Proposition 31.7 (M. Morley). *If T is totally transcendental, then $\alpha_T < \omega_1$.*

Proof. By 31.6 T is ω-stable. Corollary 19.3 provides a countable saturated model $\mathcal{A}$ of T. According to 28.1(1) $\alpha_{S\mathcal{A}} < \omega_1$. But $\alpha_{S\mathcal{A}} = \alpha_T$ by 31.3. $\square$

The absoluteness considerations alluded to at the conclusion of the proof of 31.5 lead to an effective version of 31.7: if T is totally transcendental, then α_T is an ordinal recursive in T [Sa1].

For each $\alpha < \omega_1$ there exists a totally transcendental theory T such that $\alpha_T = \alpha + 1$. An interesting example of a totally transcendental theory is the theory of differentially closed fields of characteristic 0 (DCF$_0$). It will be shown in Sec. 41 that the rank of DCF$_0$ is $\omega + 1$.

A. H. Lachlan [La2] has shown: if $\mathcal{A}$ is a model of T and $p \in S\mathcal{A}$ has a rank, then the degree of p is 1. This result is needed to prove: if T is ω-stable and not ω-categorical, then T has infinitely many countable models [La1] (cf. Proposition 35.6).

Exercise 31.8. (M. Lerman) Suppose $\mathcal{U}$ is a universal domain for T and $p \in S\mathcal{U}$ is ranked. Show p has degree 1.

Exercise 31.9. Suppose T is totally transcendental. Show α_T is not a limit ordinal.

Exercise 31.10. Suppose $\mathcal{A}$ is a field of characteristic 0. Show $\mathcal{A}$ is a universal domain for ACF$_0$ iff $\mathcal{A}$ is algebraically closed.

Exercise 31.11. T is said to have the finite basis property if for each $\mathcal{A} \in \mathcal{K}(T)$ and each set S of atomic formulas (in the language of $T \cup D\mathcal{A}$) whose sole free variable is x, there is a finite conjunction of members of S which implies (in the theory $T \cup D\mathcal{A}$) every member of S. Suppose T has the finite basis property. Show T is totally transcendental.

Section 32

Prime Model Extensions

Throughout this section T is a complete, substructure complete theory. Let $\mathcal{A} \in \mathcal{K}(T)$. $\mathcal{B}$ is a prime model extension of $\mathcal{A}$ if $\mathcal{A} \subset \mathcal{B}$, $\mathcal{B}$ is a model of T, and the following diagram can be completed as shown whenever $\mathcal{C}$ is a model of T.

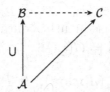

Suppose $\mathcal{B}_0, \mathcal{B}_1 \supset \mathcal{A}$. $\mathcal{B}_0$ and $\mathcal{B}_1$ are said to be isomorphic over $\mathcal{A}$ if there exists an isomorphism $f \colon \mathcal{B}_0 \xrightarrow{\approx} \mathcal{B}_1$ such that $f|A = 1_A$. It can happen that $\mathcal{A}$ has prime model extensions $\mathcal{B}_0$ and $\mathcal{B}_1$ which fail to be isomorphic over $\mathcal{A}$, but not when T is totally transcendental (Corollary 36.2). One of the results of the present section is that every $\mathcal{A} \in \mathcal{K}(T)$ has a prime model extension when T satisfies a certain condition weaker than total transcendentality. It follows from 32.12 that "every $\mathcal{A} \in \mathcal{K}(T)$ has a prime model extension" is an absolute property of T.

Proposition 32.1 (M. Morley). *Suppose $\mathcal{A} \in \mathcal{K}(T)$. (i) If $\mathcal{B} \subset \mathcal{A}$ and $\mathcal{A} \models T$, then the 1-types of $S\mathcal{B}$ realized in $\mathcal{A}$ are* *dense in $S\mathcal{B}$.* (ii) *If the 1-types of $S\mathcal{A}$ realized in $\mathcal{A}$ are dense in $S\mathcal{A}$, then $\mathcal{A} \models T$.*

Proof. Suppose $\mathcal{B} \subset \mathcal{A}$, $\mathcal{A} \models T$ and

$$N_{F(x,\underline{b})} = \{ p \mid F(x,\underline{b}) \in p \}$$

is a nonempty neighborhood of $S\mathcal{B}$. Then the completeness of $T \cup D\mathcal{B}$ implies $\mathcal{A} \models F(\underline{a},\underline{b})$ for some $a \in A$. Let $p \in S\mathcal{B}$ be the type realized in $\mathcal{A}$ by a. Then $p \in N_{F(x,\underline{b})}$.

Suppose the types of $S\mathcal{A}$ realized in $\mathcal{A}$ are dense in $S\mathcal{A}$. Let $\mathcal{C} \supset \mathcal{A}$ be a model of T. Let F be a sentence in the language of $\langle \mathcal{A}, a \rangle_{a \in A}$. An induction on the length of F demonstrates

$$\mathcal{A} \models F \text{ iff } \mathcal{C} \models F.$$

Assume F is $(Ex)G(x)$. If $\mathcal{A} \models (Ex)G(x)$, then $\mathcal{A} \models G(\underline{a})$ for some $a \in A$, and by induction $\mathcal{C} \models G(\underline{a})$. Suppose $\mathcal{C} \models (Ex)G(x)$. Then $N_{G(x)} = \{ p \mid G(x) \in p \}$ is a nonempty neighborhood of $S\mathcal{A}$, and consequently contains some p realized by some $a \in A$. But then $\mathcal{A} \models G(\underline{a})$. □

Proposition 32.2 (M. Morley). *If T is quasi-totally transcendental, then the isolated points of $S\mathcal{A}$ are dense in $S\mathcal{A}$ for every $\mathcal{A} \in \mathcal{K}(T)$.*

Proof. Let N be a nonempty neighborhood of $S\mathcal{A}$. Choose a ranked $p \in N$ so that no ranked $q \in N$ has lower rank than p. There is a neighborhood M such that $\{p\} = D^{\alpha} S\mathcal{A} \cap M$, where $\alpha = \operatorname{rank} p$. But then $\{p\} = N \cap M$. □

Theorem 32.3 (M. Morley). *If the isolated points of $S\mathcal{A}$ are dense in $S\mathcal{A}$ for every $\mathcal{A} \in \mathcal{K}(T)$, then every $\mathcal{A} \in \mathcal{K}(T)$ has a prime model extension.*

Proof. A chain $\{\mathcal{A}_\delta\}$ of structures and a sequence $\{p_\delta\}$ of 1-types are defined by induction on δ.

1. $\mathcal{A}_0 = \mathcal{A}$.
2. $\mathcal{A}_\lambda = \cup \{\mathcal{A}_\delta \mid \delta < \lambda\}$.
3. If $\mathcal{A}_\delta \models T$, then $\mathcal{A}_{\delta+1} = \mathcal{A}_\delta$.
4. Suppose $\mathcal{A}_\delta$ is not a model of T. By 32.1(ii) there is an isolated $p \in S\mathcal{A}_\delta$ not realized in $\mathcal{A}_\delta$; choose such a p to be p_δ. Let $\mathcal{A}_{\delta+1} = \mathcal{A}_\delta(a_\delta)$, where a_δ realizes p_δ.

In order to see that for some δ, $\mathcal{A}_\delta$ is a prime model extension of $\mathcal{A}$, assume $f \colon \mathcal{A} \subset \mathcal{C}$ imbeds $\mathcal{A}$ in some model $\mathcal{C}$ of T. A sequence $\{f_\delta \colon \mathcal{A}_\delta \subset \mathcal{C}\}$ is defined by induction on δ.

1*. $f_0 = f$.
2*. $f_\lambda = \cup \{f_\delta \mid \delta < \lambda\}$.
3*. If $\mathcal{A}_{\delta+1} = \mathcal{A}_\delta$, then $f_{\delta+1} = f_\delta$.
4*. Suppose $\mathcal{A}_{\delta+1} = \mathcal{A}_\delta(a_\delta)$ as in 4 above. Thus a_δ realizes $p_\delta \in S\mathcal{A}_\delta$ and p_δ is isolated. By 32.1(i) p_δ is realized by some $c \in \mathcal{C}$. Extend f_δ to $f_{\delta+1} \colon \mathcal{A}_\delta(a_\delta) \subset \mathcal{C}$ by setting $f_{\delta+1} a_\delta = c$.

Since $\mathcal{C}$ is a set and every f_δ is one-one, there must be a δ such that $f_{\delta+1} = f_\delta$. But then there must be a δ such that $\mathcal{A}_{\delta+1} = \mathcal{A}_\delta$; let γ be the least such δ. Then $\mathcal{A}_\gamma \models T$. $\mathcal{A}_\gamma$ is prime over $\mathcal{A}$, because $f_\gamma \colon \mathcal{A}_\gamma \subset \mathcal{C}$. Note that the value of γ does not depend on the choice of $\mathcal{C}$. $\qquad\square$

Corollary 32.4. *If T is quasi-totally transcendental and $\mathcal{A} \in \mathcal{K}(T)$, then $\mathcal{A}$ has a prime model extension.*

Proof. 32.2 and 32.3. $\qquad\square$

Suppose $\mathcal{A} \subset \mathcal{B} \in \mathcal{K}(T)$. $\mathcal{B}$ is an atomic extension of $\mathcal{A}$ (or atomic over $\mathcal{A}$) if for every $n > 0$ and $\langle b_1, \ldots, b_n \rangle \in B^n$, $\langle b_1, \ldots, b_n \rangle$ realizes a principal n-type of $S_n(T \cup D\mathcal{A})$.

Proposition 32.5. *If $\mathcal{C}$ is atomic over $\mathcal{B}$, and $\mathcal{B}$ is atomic over $\mathcal{A}$, then $\mathcal{C}$ is atomic over $\mathcal{A}$.*

Proof. Suppose $\langle c_1, \ldots, c_n \rangle \in C^n$ realizes a principal n-type of $S_n(T \cup D\mathcal{B})$ generated by

$$F(\underline{b}_1, \ldots, \underline{b}_m, x_1, \ldots, x_n),$$

where $\langle b_1, \ldots, b_m \rangle \in B^m$. The completeness of $T \cup D(\mathcal{A}(b_1, \ldots, b_m))$ implies that $\langle c_1, \ldots, c_n \rangle$ realizes a principal n-type of

$$S_n(T \cup D(\mathcal{A}(b_1, \ldots, b_m)))$$

generated by $F(\underline{b}_1, \ldots, \underline{b}_m, x_1, \ldots, x_n)$. Let $\langle b_1, \ldots, b_m \rangle$ realize a principal m-type of $S_m(T \cup D\mathcal{A})$ generated by $G(y_1, \ldots, y_m)$. Then $\langle c_1, \ldots, c_n \rangle$ realizes a principal n-type of $S_n(T \cup D\mathcal{A})$ generated by

$$(Ey_1) \cdots (Ey_m)[G(y_1, \ldots, y_m) \;\&\; F(y_1, \ldots, y_m, x_1, \ldots, x_n)].$$
$\square$

Theorem 32.6. *Suppose the isolated points of $S\mathcal{A}$ are dense in $S\mathcal{A}$ for every $\mathcal{A} \in \mathcal{K}(T)$. Let $\mathcal{B}$ be a prime model extension of $\mathcal{A}$. Then $\mathcal{B}$ is atomic over $\mathcal{A}$.*

Proof. Since $\mathcal{B}$ is prime over $\mathcal{A}$, it is enough to find an atomic model extension of $\mathcal{A}$. Let $\mathcal{D}$ be the model extension constructed in the proof of 32.2. Thus $\mathcal{D} = \cup\{\mathcal{D}_\delta \mid \delta < \gamma\}$; $\mathcal{D}_0 = \mathcal{A}$; $\mathcal{D}_\lambda = \cup\{\mathcal{D}_\delta \mid \delta < \lambda\}$; and $\mathcal{D}_{\delta+1} = \mathcal{D}_\delta(d_\delta)$, where d_δ realizes an isolated point of $S\mathcal{D}_\delta$. $\mathcal{D}_{\delta+1}$ is atomic over $\mathcal{D}_\delta$ since every member of $\mathcal{D}_{\delta+1}$ is of the form $t(d_\delta)$ for some term $t(x)$ in the language of $T \cup \mathcal{D}_\delta$. If follows from 32.5 and an induction on δ that $\mathcal{D}$ is atomic over $\mathcal{A}$. $\square$

Proposition 32.7. *Suppose $\mathcal{C} \supset \mathcal{B} \supset \mathcal{A}$. If $\mathcal{C}$ is atomic over $\mathcal{A}$, and $\mathcal{B}$ is finitely generated over $\mathcal{A}$, then $\mathcal{C}$ is atomic over $\mathcal{B}$.*

Proof. Suppose $\mathcal{B} = \mathcal{A}(b_1, \ldots, b_n)$ and $\mathcal{C} = \mathcal{B}(c_1, \ldots, c_m)$. The proposition is proved by induction on n, so it is safe to assume $\mathcal{B} = \mathcal{A}(b)$. For a fixed $\mathcal{B}$ the proposition proceeds by induction on m thanks to 32.5, so assume $\mathcal{C} = \mathcal{B}(c)$.

Let $F(x, y)$ generate the principal 2-type of $S_2(T \cup D\mathcal{A})$ realized by $\langle b, c \rangle$. Let $q \in S_1(T \cup D\mathcal{B})$ be the 1-type realized by c. Clearly $F(\underline{b}, y) \in q$. To see that $F(\underline{b}, y)$ generates q, suppose

$$T \cup DC \vdash H(\underline{c}),$$

where $H(y)$ is a formula in the language of $T \cup D\mathcal{B}$. Then $H(y) = H(\underline{b}, y)$, where $H(x, y)$ is a formula in the language of $T \cup D\mathcal{A}$, and

$$T \cup D\mathcal{A} \vdash F(x, y) \to H(x, y),$$
$$T \cup D\mathcal{B} \vdash F(\underline{b}, y) \to H(y).$$

$\square$

Let $\mathcal{C}$ be a prime model extension of $\mathcal{A}$. Suppose $\mathcal{C} = \cup \{\mathcal{C}_\delta \mid \delta < \gamma\}$, where $\mathcal{C}_0 = \mathcal{A}$, $\mathcal{C}_\lambda = \cup \{\mathcal{C}_\delta \mid \delta < \lambda\}$, $\mathcal{C}_{\delta+1} = \mathcal{C}_\delta(c_\delta)$, and c_δ realizes a principal 1-type of $S\mathcal{C}_\delta$. Then $\mathcal{C}$ is said to be a Morley prime extension of $\mathcal{A}$ (or Morley prime over $\mathcal{A}$). The prime model extension of 32.3 is Morley.

Proposition 32.8. *Suppose* $\mathcal{C} \supset \mathcal{B} \supset \mathcal{A}$. *If* $\mathcal{C}$ *is Morley prime over* $\mathcal{A}$, *and* $\mathcal{B}$ *is finitely generated over* $\mathcal{A}$, *then* $\mathcal{C}$ *is Morley prime over* $\mathcal{B}$.

Proof. By hypothesis $\mathcal{C} = \cup \{\mathcal{C}_\delta \mid \delta < \gamma\}$, where $\mathcal{C}_0 = \mathcal{A}$, $\mathcal{C}_\lambda = \cup \{\mathcal{C}_\delta \mid \delta < \lambda\}$, $\mathcal{C}_{\delta+1} = \mathcal{C}_\delta(c_\delta)$, and c_δ realizes a principal 1-type of $S\mathcal{C}_\delta$. Define:

$$\mathcal{C}_0^* = \mathcal{B},$$
$$\mathcal{C}_\lambda^* = \cup \{\mathcal{C}_\delta^* \mid \delta < \lambda\},$$
$$\mathcal{C}_{\delta+1}^* = \mathcal{C}_\delta^*(c_\delta).$$

Clearly $\mathcal{C} = \cup \{\mathcal{C}_\delta^* \mid \delta < \gamma\}$. In order to see $\mathcal{C}_{\delta+1}^*$ is atomic over $\mathcal{C}_0^*$, choose $\rho < \gamma$ so that $\mathcal{B} \subset \mathcal{C}_\rho$. If $\delta \geq \rho$, then $\mathcal{C}_\delta^* = \mathcal{C}_\delta$ and so $\mathcal{C}_{\delta+1}^*$ is atomic over $\mathcal{C}_\delta^*$. Assume $\delta < \rho$. Then $\mathcal{C}_\rho^* = \mathcal{C}_\rho$ and so is atomic over $\mathcal{C}_\delta$ by 32.5. But then $\mathcal{C}_\rho^*$, and hence $\mathcal{C}_{\delta+1}^*$, is atomic over $\mathcal{C}_\delta^*$ by 32.7. $\square$

Suppose $\mathcal{C} \supset \mathcal{B} \supset \mathcal{A}$ and $\mathcal{C}$ is atomic over $\mathcal{A}$. It can happen that $\mathcal{C}$ is not atomic over $\mathcal{B}$. In that event $\mathcal{B}$ is not finitely generated over $\mathcal{A}$ according to 32.7. The next result supplies a condition sufficient for $\mathcal{C}$ to be atomic over $\mathcal{B}$. It is needed for the proof of Shelah's uniqueness theorem in Sec. 36, and was first proved by Shelah for totally transcendental theories.

Suppose $\mathcal{A} \subset \mathcal{B} \subset \mathcal{C} \in \mathcal{K}(T)$; $\mathcal{B}$ is normal over $\mathcal{A}$ in $\mathcal{C}$ if for all $p \in S\mathcal{A}$, all or none of the realizations of p in $\mathcal{C}$ belong to $\mathcal{B}$.

Lemma 32.9 (L. Harrington). *Suppose the isolated points of $S\mathcal{A}$ are dense in $S\mathcal{A}$ for every $\mathcal{A} \in \mathcal{K}(T)$. If $\mathcal{C}$ is an atomic model extension of $\mathcal{A}$, and $\mathcal{B}$ is normal over $\mathcal{A}$ in $\mathcal{C}$, then $\mathcal{C}$ is atomic over $\mathcal{B}$.*

Proof. Suppose for a reductio ad absurdum that $q \in S_n(T \cup D\mathcal{B})$ is a limit point and is realized in $\mathcal{C}$. For notational simplicity allow $n = 1$. Choose $c \in C$ to realize q. The following Skolem–Löwenheim argument reduces $\mathcal{C}$, $\mathcal{B}$ and $\mathcal{A}$ to the countable level, where the methods of Sec. 21 can be applied. $\mathcal{A}_n, \mathcal{B}_n$ and $\mathcal{C}_n$ are defined by induction on n.

1. Let $F(x, \underline{a}_1, \ldots, \underline{a}_m)$ generate the principal 1-type p realized by c over $\mathcal{A}$. Let $\mathcal{A}_0$ be the least substructure of $\mathcal{A}$ generated by $a_1, \ldots, a_m$. Then c realizes an isolated point of $S\mathcal{A}^*$ generated by $F(x, \underline{a}_1, \ldots, \underline{a}_m)$ whenever $\mathcal{A}_0 \subset \mathcal{A}^* \subset \mathcal{A}$. Clearly q is a pre-image of p; i.e. $p = (Si)q$, where i: $\mathcal{A} \subset \mathcal{B}$ is the inclusion map.
2. Extend $\mathcal{A}_0$ to a countable $\mathcal{B}_0 \subset \mathcal{B}$ so that p has more than one pre-image in $S\mathcal{B}_0$. If no such $\mathcal{B}_0$ existed, then q would be an isolated point of $S\mathcal{B}$ by 26.2.
3. Extend $\mathcal{B}_0(c)$ to a countable $\mathcal{C}_0 \prec \mathcal{C}$.
4. Fix $n > 0$. Assume $\mathcal{A}_{n-1} \subset \mathcal{A}$, $\mathcal{A}_{n-1} \subset \mathcal{B}_{n-1} \subset \mathcal{B}$, $\mathcal{B}_{n-1}(c) \subset \mathcal{C}_{n-1} \prec \mathcal{C}$, and $\mathcal{C}$ is countable. As in step 1, extend

$\mathcal{A}_{n-1}$ to a countable $\mathcal{A}_n \subset \mathcal{A}$ so that for each k and each $\langle d_1, \ldots, d_k \rangle \in C^k_{n-1}$, the isolated point of $S_k(T \cup D\mathcal{A})$ realized by $\langle d_1, \ldots, d_k \rangle$ is generated by a formula whose individual constants belong to $\mathcal{A}_n$. The formula exists because $\mathcal{C}$ is atomic over $\mathcal{A}$.

5. Extend $\mathcal{A}_n \cup \mathcal{B}_{n-1}$ to a countable $\mathcal{B}^*_{n-1} \subset \mathcal{B}$ so that: if d_1, $d_2 \in C_{n-1}$ realize the same type over $\mathcal{A}_n$ and $d_1 \in \mathcal{B}_{n-1}$, then $d_2 \in \mathcal{B}^*_{n-1}$. To see that $\mathcal{B}^*_{n-1}$ exists, it suffices to see that the conditions imposed on d_1 and d_2 imply $d_2 \in B$. $\mathcal{A}_n$ was chosen in step 4 in such a manner that d_1 and d_2 realize the same principal type over $\mathcal{A}$ by virtue of realizing the same type over $\mathcal{A}_n$. So d_2 belongs to $\mathcal{B}$, since $\mathcal{B}$ is normal over $\mathcal{A}$ in $\mathcal{C}$.

6. As in step 2, extend $\mathcal{B}^*_{n-1}$ to a countable $\mathcal{B}_n \subset \mathcal{B}$ so that p^*, the type of c over $\mathcal{B}^*_{n-1}$, splits in $S\mathcal{B}_n$. Extend $\mathcal{B}_n \cup \mathcal{C}_{n-1}$ to a countable $\mathcal{C}_n \prec \mathcal{C}$.

Let $\mathcal{A}_\infty = \cup\{\mathcal{A}_n\}$, $\mathcal{B}_\infty = \cup\{\mathcal{B}_n\}$ and $\mathcal{C}_\infty = \cup\{\mathcal{C}_n\}$. Then $\mathcal{C}_\infty \supset \mathcal{B}_\infty \supset \mathcal{A}_\infty$; $\mathcal{C}_\infty$ is an atomic model extension of $\mathcal{A}_\infty$ by step 4; $\mathcal{B}_\infty$ is normal over $\mathcal{A}_\infty$ in $\mathcal{C}_\infty$ by step 5; and $c \in C_\infty$ does not realize an isolated point of $S\mathcal{B}_\infty$ by step 6.

In order to contradict the last assertion about c, let $r \in S\mathcal{A}_\infty$ be the type realized by c. Since r is isolated, and since the isolated points of $S\mathcal{B}_\infty$ are dense in $S\mathcal{B}_\infty$, r has an isolated pre-image $r^* \in S\mathcal{B}_\infty$. $\mathcal{C}_\infty \models T$, so some $c^* \in C_\infty$ realizes r^* by 32.1(i). The countability of $\mathcal{C}_\infty$, 21.2 and 21.4 imply that $\mathcal{C}_\infty$ is a homogeneous model of $T \cup D\mathcal{A}_\infty$. Let $f: \mathcal{C}_\infty \xrightarrow{\approx} \mathcal{C}_\infty$ be an automorphism such that $fc = c^*$ and $f \mid \mathcal{A}_\infty = 1_{\mathcal{A}_\infty}$. Since $\mathcal{B}_\infty$ is normal over $\mathcal{A}_\infty$ in $\mathcal{C}_\infty$, $f[\mathcal{B}_\infty] = \mathcal{B}_\infty$. But then c realizes an isolated point of $S\mathcal{B}_\infty$ because c^* does. $\quad\square$

Suppose $\mathcal{A} \subset \mathcal{B}$ and $\mathcal{B}$ is a model of T. $\mathcal{B}$ is a minimal model extension of $\mathcal{A}$ (or minimal over $\mathcal{A}$) if there is no $\mathcal{C} \subset \mathcal{B}$ such that $\mathcal{A} \subset \mathcal{C}$, $\mathcal{C}$ is a model of T and $\mathcal{C} \neq \mathcal{B}$. It can happen that $\mathcal{C}$ has a prime model extension which fails to be minimal.

Proposition 32.10. *Suppose the points of $S\mathcal{A}$ of (Morley) rank 0 are dense in $S\mathcal{A}$ for every $\mathcal{A} \in \mathcal{K}(T)$. Then every $\mathcal{A} \in \mathcal{K}(T)$ has a minimal prime model extension.*

Proof. Let $\mathcal{B} \supset \mathcal{A}$; $\mathcal{B}$ is algebraic over $\mathcal{A}$ if every $b \in \mathcal{B}$ realizes a point of $S\mathcal{A}$ of (Morley) rank 0. If $\mathcal{C}$ is algebraic over $\mathcal{B}$, and $\mathcal{B}$ is algebraic over $\mathcal{A}$, then $\mathcal{C}$ is algebraic over $\mathcal{A}$. The last assertion follows from 32.13. Recall the construction of the prime model extension of $\mathcal{A}$ in 32.3: $p_\delta \in S\mathcal{A}_\delta$ must have (Morley) rank 0. So $\mathcal{A}$ has a prime model extension $\mathcal{B}$ algebraic over $\mathcal{A}$. Suppose $\mathcal{A} \subset \mathcal{C} \subset \mathcal{B}$ and $\mathcal{C}$ is a model of T. Clearly $\mathcal{C}$ is algebraic over $\mathcal{A}$. $\mathcal{C}$ is algebraically closed if every $p \in S\mathcal{C}$ of Morley rank 0 is realized in $\mathcal{C}$. By 32.1 $\mathcal{C}$ is algebraically closed. It follows that every algebraic extension of $\mathcal{C}$, in particular $\mathcal{B}$, equals $\mathcal{C}$. $\square$

Exercise 32.11. Let $\mathcal{B}$ be a model of T, and $J(x)$ be a formula in the language of $T \cup D\mathcal{B}$. Then the set of realizations of $J(x)$ in $\mathcal{B}$ is finite iff $\{q \mid J(x) \in q \in S\mathcal{B}\}$ is finite and every member of $\{q \mid J(x) \in q \in S\mathcal{B}\}$ has rank 0.

Exercise 32.12. Suppose every countable $\mathcal{A} \in \mathcal{K}(T)$ has a prime model extension. Show every $\mathcal{A} \in \mathcal{K}(T)$ has a prime model extension.

Exercise 32.13. Show (i) is equivalent to (ii).

(i) b is algebraic over $\mathcal{A}$.
(ii) There exists a formula $F(x)$ (in the language of $T \cup D\mathcal{A}$) such that b realizes $F(x)$ and such that the number of realizations of $F(x)$ in any model extension of $\mathcal{A}$ is finite.

Exercise 32.14 (J. Ressayre). Suppose $\mathcal{B}$ and $\mathcal{C}$ are Morley prime extensions of $\mathcal{A}$. Show $\mathcal{B}$ and $\mathcal{C}$ are isomorphic over $\mathcal{A}$.

Section 33

Prime Extensions of Wellordered Chains

Morley created Theorem 33.1 in order to omit a type, but it turned out that Corollary 32.4 sufficed. It seems likely that 33.1 will have an essential use someday. If not, its existence is justified by its internal beauty.

T is complete and substructure complete. Suppose $\{\mathcal{A}_\delta \mid \delta < \alpha\}$ is a wellordered chain of members of $\mathcal{K}(T)$; i.e. $\mathcal{A}_\gamma \subset \mathcal{A}_\delta$ when $\gamma < \delta$. A model extension of $\{\mathcal{A}_\delta \mid \delta < \alpha\}$ is a chain $\{\mathcal{B}_\delta \mid \delta < \alpha\}$ of models of T that satisfies the below commutative diagram. $\{\mathcal{B}_\delta \mid \delta < \alpha\}$ is a *prime* model extension if $\mathcal{B}_\delta$ is a *prime* model extension of $\mathcal{A}_\delta$ for all $\delta < \alpha$.

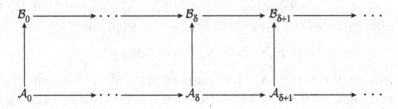

Theorem 33.1 (M. Morley). *If T is quasi-totally transcendental, then every wellordered chain of substructures of models of T has a prime model extension.*

Proof. Let $\{\mathcal{A}_\delta \,|\, \delta < \alpha\}$ be a chain of members of $\mathcal{K}(T)$. Assume $\mathcal{A}_\lambda = \cup \{\mathcal{A}_\delta \,|\, \delta < \lambda\}$ for every limit ordinal $\lambda < \alpha$; if this is not the case, insert additional structures in the chain to make it so. Assume α is a limit ordinal, and define $\mathcal{A}_\alpha = \cup \{\mathcal{A}_\delta \,|\, \delta < \alpha\}$. A sequence $\{\mathcal{A}_\delta^\beta \,|\, \delta \le \alpha\}$ of chains is defined by induction on β.

1. $\mathcal{A}_\delta^0 = \mathcal{A}_\delta$.
2. $\mathcal{A}_\delta^\lambda = \cup \{\mathcal{A}_\delta^\beta \,|\, \beta < \lambda\}$.
3. If $\mathcal{A}_\delta^\beta \models T$, then $\mathcal{A}_\delta^{\beta+1} = \mathcal{A}_\delta^\beta$.
4. Suppose $\mathcal{A}_\delta^\beta$ is not a model of T, but $\mathcal{A}_\gamma^\beta \models T$ for all $\gamma < \delta$. A sequence $\{p_\gamma^\beta \,|\, \delta \le \gamma \le \alpha\}$ of isolated points is defined by induction on γ. By 32.1 and 32.2 some isolated $p \in S\mathcal{A}_\delta^\beta$ is not realized in $\mathcal{A}_\delta^\beta$; let p_δ^β be such a p. Fix $\gamma \le \alpha$. Suppose p_γ^β is an isolated point of $S\mathcal{A}_\gamma^\beta$. Then the set of all pre-images of p_γ^β in $S\mathcal{A}_{\gamma+1}^\beta$ is open; let $p_{\gamma+1}^\beta$ be a pre-image of the lowest possible rank and of the lowest degree in that rank. As in the proof of 32.2, $p_{\gamma+1}^\beta$ is an isolated point of $S\mathcal{A}_{\gamma+1}^\beta$.

   Fix $\lambda \le \alpha$. Suppose for all $\gamma < \lambda$, $p_{\gamma+1}^\beta$ is a pre-image of p_γ^β of the lowest possible rank and of the lowest possible degree in that rank. Suppose further that p_γ^β is an isolated point of $S\mathcal{A}_\sigma^\beta$ for all $\gamma < \lambda$, and that p_σ^β is a pre-image of p_τ when $\tau < \sigma < \lambda$. Define $p_\lambda^\beta = \cup \{p_\gamma^\beta \,|\, \gamma < \lambda\}$. To see that p_λ^β is isolated, observe that there is a $\mu < \lambda$ such that

$$\text{rank } p_\gamma^\beta = \text{rank } p_\lambda^\beta \quad \text{and} \quad \deg p_\gamma^\beta = p_\lambda^\beta$$

whenever $\mu \le \gamma \le \lambda$. The existence of μ follows from the rank and degree rules (29.3 and 29.5): as γ approaches λ, the rank of p_γ^β can decrease only finitely often, and so must be constant from some point on; after that the degree can decrease only finitely often. If $\mu \le \gamma < \lambda$, then p_γ^β is the sole pre-image of p_μ^β in $S\mathcal{A}_\gamma^\beta$, since any other pre-image would have an impossibly lower rank or degree. It follows from 26.2 that p_λ^β is isolated in $S\mathcal{A}_\lambda^\beta$.

5. Define δ as in clause 4. Suppose for some $\gamma > \delta$, p_γ^β is realized in $\mathcal{A}_\gamma^\beta$. Let γ^0 be the least such γ. Choose $a \in A_{\gamma^0}^\beta$ to realize $p_{\gamma^0}^\beta$. Define $\mathcal{A}_\gamma^{\beta+1} = \mathcal{A}_\lambda^\beta(a)$ for all $\lambda \geq \delta$, and $\mathcal{A}_\gamma^{\beta+1} = \mathcal{A}_\gamma^\beta$ for all $\gamma < \delta$.

6. Suppose for every $\gamma > \delta$, p_γ^β is not realized in $\mathcal{A}_\gamma^\beta$. Choose an a such that a realizes p_α^β. Define $\mathcal{A}_\gamma^{\beta+1} = \mathcal{A}_\gamma^\beta(a)$ for all $\gamma \geq \delta$, and $\mathcal{A}_\gamma^{\beta+1} = \mathcal{A}_\gamma^\beta$ for all $\gamma < \delta$.

Note that in steps 5 and 6 above, a realizes an isolated point of $S\mathcal{A}_\gamma^\beta$, namely p_γ^β, for all $\gamma \geq \delta$. Consequently, the argument of 32.3 shows $\bigcup_\beta \mathcal{A}_\gamma^\beta$ is a Morley prime extension of $\mathcal{A}_\gamma$ for all $\gamma \leq \alpha$. $\qquad\square$

It is not known if the hypothesis of 33.1 can be replaced by the weaker hypothesis of 32.3.

Section 34

Order Indiscernibles

Let T be a theory with an infinite model. A seemingly innocent question about T arising from quotidian experiences is: does T have an infinite model with a nontrivial automorphism? The answer, found by Ehrenfeuct and Mostowski, is yes. Their argument was based on the notion of indiscernibility. Later Morley discovered that indiscernibles could be used to omit a type, as in the proof of 37.2. They can also be used to construct isomorphisms, as in the proof of 36.1, and to define a notion of dimension for models of ω_1-categorical theories, as in Secs. 38 and 39. In this section they are needed to show that every theory categorical in some uncountable power is ω-stable.

Let $\langle I, < \rangle$ be a linear ordering, and let $\mathcal{A}$ be a structure such that $I \subset A$. The elements of I are order indiscernible in $\mathcal{A}$ if

$$\mathcal{A} \models F(\underline{i}_1, \ldots, \underline{i}_n) \leftrightarrow F(\underline{i}'_1, \ldots, \underline{i}'_n)$$

whenever $i_1 < \cdots < i_n$ and $i'_1 < \cdots < i'_n$ are ordered n-tuples of $\langle I, < \rangle$ and $F(x_1, \ldots, x_n)$ is a formula in the language underlying $\mathcal{A}$.

Theorem 34.1 (A. Ehrenfeuct, A. Mostowski). *Let T be a theory with an infinite model and let $\langle I, < \rangle$ be a linear ordering.*

Then there exists a model $\mathcal{A}$ of T such that the elements of I are order indiscernible in $\mathcal{A}$.

Proof. Extend the language of T by adding an individual constant $\underline{i}$ for each $i \in I$. Extend T to T^* by adding:

(a) $\{\underline{i} \neq \underline{i}' \mid i, i' \in I; i \neq i'\}$;
(b) every sentence of the form

$$F(\underline{i}_1, \ldots, \underline{i}_n) \leftrightarrow F(\underline{i}'_1, \ldots, \underline{i}'_n),$$

where $i_1 < \cdots < i_n$, $i'_1 < \cdots < i'_n$, and $F(x_1, \ldots, x_n)$ is a formula in the language of T.

Any model of T^* can serve as $\mathcal{A}$. To see that T^* is consistent, let $\mathcal{B}$ be an infinite model of T. A celebrated combinatoric result of Ramsey is needed to show that every finite subset of $T^* - T$ is satisfiable in $\mathcal{B}$.

For any set G, let $G^{(n)}$ be the set of all n-element subsets of G.

Ramsey's Theorem. *Let G be infinite, and let $G^{(n)} = H_1 \cup \cdots \cup H_m$ be a partition of $G^{(n)}$ into m mutually disjoint sets. Then there exists an infinite $K \subset G$ and a j such that $K^{(n)} \subset H_j$.*

A typical finite subset of $T^* - T$ is:

$$(b1)\, F_1(\underline{i}_{1,1}, \ldots, \underline{i}_{1,n_1}) \leftrightarrow F_1(\underline{i}'_{1,1}, \ldots, \underline{i}'_{1,n_1});$$

$$\vdots$$

$$(bd)\, F_d(\underline{i}_{d,1}, \ldots, \underline{i}_{d,n_d}) \leftrightarrow F_d(\underline{i}'_{d,1}, \ldots, \underline{i}'_{d,n_d}).$$

Let $<^*$ be an arbitrary linear ordering of B. An initial use of Ramsey's theorem furnishes an infinite $B_1 \subset B$ such that

$$\mathcal{B} \models F_1(\underline{b}_1, \ldots, \underline{b}_{n_1}) \leftrightarrow F_1(\underline{b}'_1, \ldots, \underline{b}'_{n_1})$$

whenever $b_1 <^* \cdots <^* b_{n_1}$ and $b'_1 <^* \cdots <^* b'_{n_1}$ are ordered n_1-tuples of $\langle B_1, <^* \rangle$. A second use of Ramsey's theorem affords an infinite $B_2 \subset B_1$ such that

$$\mathcal{B} \models F_2(\underline{b}_1, \ldots, \underline{b}_{n_2}) \leftrightarrow F_2(\underline{b}'_1, \ldots, \underline{b}'_{n_2})$$

whenever $b_1 <^* \cdots <^* b_{n_2}$ and $b'_1 <^* \cdots <^* b'_{n_2}$ are ordered n_2-tuples of $\langle B_2, <^* \rangle$. d consecutive applications of Ramsey's theorem end in an infinite $B_d \subset \cdots \subset B_1 \subset B$ such that $\mathcal{B}$, with $\langle B_d, <^* \rangle$ playing the part of $\langle I, < \rangle$, is a model of T, $(a), (b1), \ldots, (bd)$. □

Theorem 34.2 (A. Ehrenfeuct, A. Mostowski). *Let T be a theory with an infinite model, and let $\langle I, < \rangle$ be a linear ordering. Then there exists a model $\mathcal{B}$ of T such that $I \subset B$ and every endomorphism (respectively automorphism) of $\langle I, < \rangle$ can be extended to an elementary endomorphism (respectively automorphism) of $\mathcal{B}$.*

Proof. Let $T^{\mathcal{S}}$ be the Skolemization of T as in Sec. 11. By 34.1 there exists a model $\mathcal{A}^{\mathcal{S}}$ of $T^{\mathcal{S}}$ such that the elements of I are order indiscernible in $\mathcal{A}^{\mathcal{S}}$. Let $\mathcal{B}^{\mathcal{S}}$ be the Skolem hull of I in $\mathcal{A}^{\mathcal{S}}$. Then $\mathcal{B}$ is the desired model of T. The elements of I are order indiscernible in $\mathcal{B}^{\mathcal{S}}$ since $\mathcal{B}^{\mathcal{S}} \prec \mathcal{A}^{\mathcal{S}}$. Suppose $f\colon \langle I, < \rangle \to \langle I, < \rangle$ is an endomorphism. A typic member b of $\mathcal{B}^{\mathcal{S}}$ can be represented as $t(i_1, \ldots, i_n)$, where t is an n-place Skolem function of $\mathcal{A}^{\mathcal{S}}$ and $i_1 < \cdots < i_n$ is an n-tuple of $\langle I, < \rangle$. The natural extension of f from I to $\mathcal{B}^{\mathcal{S}}$ is given by

$$\tilde{f}b = \tilde{f}(t(i_1, \ldots, i_n)) = t(fi_1, \ldots, fi_n).$$

The value of $\tilde{f}b$ does not depend on the choice of representation of b, because I is order indiscernible in $\mathcal{B}^{\mathcal{S}}$ and f is an order endomorphism of I: if $t_1(i_1) = t_2(i_2)$ and $i_1 < i_2$, then $fi_1 < fi_2$ and $t_1(fi_1) = t_2(fi_2)$. On similar grounds $\tilde{f}$ is an endomorphism. $\tilde{f}$ is elementary since $T^{\mathcal{S}}$ is model complete. If f maps I onto I, then $\tilde{f}$ maps $\mathcal{B}^{\mathcal{S}}$ onto $\mathcal{B}^{\mathcal{S}}$. □

If T is totally transcendental, then Exercise 35.11 yields for every infinite cardinal κ, a model $\mathcal{B}$ of T and a set $I \subset B$ such that card $I =$ card $\mathcal{B} = \kappa$ and every one-one map of I *into* I can be extended to an automorphism of $\mathcal{B}$.

Theorem 34.3 (M. Morley). *Let T be a theory with an infinite model, and let κ be an infinite cardinal. Then there exists a model $\mathcal{B}$ of T of cardinality κ such that for each countable $Y \subset B$, only countably many 1-types of $S_1 T(\langle \mathcal{B}, y\rangle_{y \in Y})$ are realized in $\mathcal{B}$.*

Proof. The "ω-stability" of $\mathcal{B}$ over every countable $Y \subset B$ will follow from the fact there are only countably many cuts in any countable wellordering. (A cut in a linear ordering $\langle J, < \rangle$ is a set $K \subset J$ such that $x \in K$ and $y \in J - K$ imply $x < y$.)

Let $\langle I, < \rangle$ be the wellordering of the ordinals less than κ. Suppose I, $\mathcal{A}^{\mathcal{S}}$ and $\mathcal{B}^{\mathcal{S}}$ are as they were in the proof of 34.2. Let Y be a countable subset of $\mathcal{B}^{\mathcal{S}}$. It is safe to assume Y is the Skolem hull of some countable $J \subset I$, since otherwise Y can be augmented. A typic member of $\mathcal{B}^{\mathcal{S}}$ is $t(i_1, \ldots, i_n)$, where t is an n-place Skolem function of $\mathcal{A}^{\mathcal{S}}$ and $i_1 < \cdots < i_n$ is an n-tuple of $\langle I, < \rangle$. The 1-type of $S_1 T(\langle \mathcal{B}^{\mathcal{S}}, y\rangle_{y \in Y})$ realized by $t(i_1, \ldots, i_n)$ is determined by t and the positions of $i_1, \ldots, i_n$ in $\langle I, < \rangle$ relative to J, since the member of I are order indiscernible in $\mathcal{B}^{\mathcal{S}}$. For instance suppose: $y = sj$ for some Skolem function s and some $j \in J$; $i < j$; and $i' < j$. Then

$$\mathcal{B}^{\mathcal{S}} \models F(\underline{y}, ti) \leftrightarrow F(\underline{y}, ti').$$

There are only countably many t's. Choosing positions for $i_1, \ldots, i_n$ relative to J amounts to choosing finitely many cuts in J. □

Corollary 34.4 (M. Morley). *If T is categorical in some uncountable power, then T is ω-stable.*

Proof. Suppose $\mathcal{A}$ is a countable model of T such that $S_1 T(\langle \mathcal{A}, a\rangle_{a \in A})$ is uncountable. It follows from 10.3 and ω_1 uses of 15.1 that there exists a $\mathcal{C} \succ \mathcal{A}$ such that card $\mathcal{C} = \omega_1$ and uncountably many members of $S_1 T(\langle \mathcal{A}, a\rangle_{a \in A})$ are realized in $\mathcal{C}$. Let κ be an uncountable cardinal such that all models of T of cardinality κ are isomorphic. Elementarily extend $\mathcal{C}$ to $\mathcal{D}$, a structure of cardinality κ. Then $\mathcal{D}$ realizes uncountably many types over $\mathcal{A}$. But that is not possible, since $\mathcal{D}$ is isomorphic to the $\mathcal{B}$ supplied by 34.3. $\qquad\square$

Corollary 34.5 (M. Morley). *Suppose $\kappa > \omega$ and T is κ-categorical. Then every model of T of cardinality κ is saturated.*

Proof. By 34.4 T is ω-stable. Let $\mathcal{A}$ be a model of T of cardinality κ. If κ is regular then $\mathcal{A}$ is saturated by 19.3. Suppose κ is singular and ρ is a regular cardinal less than κ. By 19.2 there is a ρ-saturated $\mathcal{B} \succ \mathcal{A}$ of cardinality κ. But $\mathcal{B} \approx \mathcal{A}$, so $\mathcal{A}$ is ρ-saturated. $\qquad\square$

Exercise 34.6 (M. Morley). Suppose $\kappa > \omega$ and every model of T of cardinality κ is ω_1-saturated. Show T is ω-stable.

Section 35

Indiscernibles and ω-Stability

The elements of I are indiscernible in $\mathcal{A}$ if $I \subset A$ and

$$\mathcal{A} \models F(\underline{i}_1, \ldots, \underline{i}_n) \leftrightarrow F(\underline{i}'_1, \ldots, \underline{i}'_n)$$

whenever $\{i_1, \ldots, i_n\}$ and $\{i'_1, \ldots, i'_n\}$ are n-element subsets of I and $F(x_1, \ldots, x_n)$ is a formula in the language underlying $\mathcal{A}$. Suppose $\mathcal{A}$ is an uncountable model of an ω-stable theory. Then for each regular $\kappa \leq \operatorname{card} \mathcal{A}$, $\mathcal{A}$ contains a set of indiscernibles of cardinality κ according to 35.7 and 31.6. If $\mathcal{A}$ is a prime model extension of $\mathcal{B}$, then every set of indiscernibles in $\langle \mathcal{A}, b \rangle_{b \in B}$ is countable by 35.9. The proofs of 35.9 and 35.10 demonstrate the power of the rank and degree machinery of Sec. 29.

Theorem 35.1 (M. Morley). *Let I be an infinite set of order indiscernibles in $\mathcal{A}$. If $T\mathcal{A}$ is ω-stable, then I is a set of indiscernibles in $\mathcal{A}$.*

Proof. P_n is the group of all permutations of $\{1, \ldots, n\}$. Fix $F(x_1, \ldots, x_n)$ and $i_1 < \cdots < i_n \in I$. Let P_n^+ be the set of all $s \in P_n$ such that

$$\mathcal{A} \models F(\underline{i}_{s1}, \ldots, \underline{i}_{sn}).$$

If $P_n^+ = P_n$ or $P_n^+ = 0$, then the theorem is proved. Suppose $s \in P_n^+$ and $t \in P_n - P_n^+$. s and t differ by a product of transpositions. (The transposition $(k, k + 1)$) permutes $\{1, \ldots, k - 1, k, k + 1, k + 2, \ldots, n\}$ to $\{1, \ldots, k - 1, k + 1, k, k + 2, \ldots, n\}$.) It follows there exist $s \in P_n^+$ and $t \in P_n - P_n^+$ such that

$$t = (k, k + 1) \cdot s$$

for some k. Let $G(x_1, \ldots, x_n)$ be $F(x_{s1}, \ldots, x_{sn})$. Then

(1) $\mathcal{A} \models G(\underline{i}_1, \ldots, \underline{i}_n)$
(2) $\mathcal{A} \models \sim G(\underline{i}_1, \ldots, \underline{i}_{k-1}, \underline{i}_{k+1}, \underline{i}_k, \underline{i}_{k+2}, \ldots, \underline{i}_n)$.

Note that (1) and (2) hold when $i_1 < \cdots < i_n$ is replaced by $i'_1 < \cdots < i'_n \in I$, since I is a set of order indiscernibles in $\mathcal{A}$. Let $\langle R, < \rangle$ be the standard ordering of the real numbers. It follows from the compactness theorem that there exists a $\mathcal{B} \equiv \mathcal{A}$ such that $R \subset B$ and

$$\mathcal{B} \models G(\underline{r}_1, \ldots, \underline{r}_n)$$
$$\mathcal{B} \models \sim G(\underline{r}_1, \ldots, \underline{r}_{k-1}, \underline{r}_{k+1}, \underline{r}_k, \underline{r}_{k+2}, \ldots, \underline{r}_n)$$

whenever $r_1 < \cdots < r_n \in R$. The use of compactness is legitimate because I is infinite.

Suppose $r < r' \in R$. There exist rationals $\{q_i \mid 1 \leq i \leq n \text{ and } i \neq k\}$ such that

$$q_1 < \cdots < q_{k-1} < r < q_{k+1} < r' < q_{k+2} < \cdots < q_n.$$

Consequently,

$$\mathcal{B} \models G(\underline{q}_1, \ldots, \underline{q}_{k-1}, \underline{r}, \underline{q}_{k+1}, \ldots, \underline{q}_n)$$
$$\mathcal{B} \models \sim G(\underline{q}_1, \ldots, \underline{q}_{k-1}, \underline{r}', \underline{q}_{k+1}, \ldots, \underline{q}_n).$$

Thus distinct members of R realize distinct 1-types over the rationals Q in $\mathcal{B}$. But then $T\mathcal{A}$ is not ω-stable. $\qquad \square$

Suppose A is a set, R is an n-place relation on A and $I \subset A$. R is said to be connected (antisymmetric) over I if for each

n-element subset $\{i_1, \ldots, i_n\}$ of I there is a permutation s of $\{1, \ldots, n\}$ such that $R(\underline{i}_{s1}, \ldots, \underline{i}_{sn})$ holds (does not hold).

R is a definable relation of $\mathcal{A}$ if there exists a formula $F(x_1, \ldots, x_n)$ in the language of $T\mathcal{A}$ such that

$$R(a_1, \ldots, a_n) \quad \text{iff} \mathcal{A} \models F(\underline{a}_1, \ldots, \underline{a}_n)$$

for all $\langle a_1, \ldots, a_n \rangle \in A^n$.

Theorem 35.2. *If $T\mathcal{A}$ is ω-stable, then no $\mathcal{A}$-definable relation is both connected and antisymmetric over any infinite $I \subset A$.*

Proof. Similar to 35.1. Suppose $F(x_1, \ldots, x_n)$ defines a relation in $\mathcal{A}$ which is connected and antisymmetric over some infinite $I \subset A$. Let $<$ be any ordering of I. As in the proof of 34.1, finitely many consecutive applications of Ramsey's theorem yield an infinite $J \subset I$ such that J is a set of order indiscernibles in $\mathcal{A}$ with respect to all formulas obtained by permuting the variables of $F(x_1, \ldots, x_n)$. Since the relation defined by $F(x_1, \ldots, x_n)$ is connected and antisymmetric over J, the elements of J are not indiscernible with respect to $F(x_1, \ldots, x_n)$. Now proceed as in 35.1. $\square$

S. Shelah [Sh1] has proved the converse of a slight modification of 35.2.

Corollary 35.3. *If $\mathcal{A}$ has an infinite, definable linear ordering, then $T\mathcal{A}$ is not ω-stable.*

Let $\langle I, < \rangle$ be a linear ordering, and let $\mathcal{A}$ be a structure such that $I \subset A$. Suppose $Y \subset A$. The elements of I are order indiscernible over Y in $\mathcal{A}$ if they are order indiscernible in $\langle \mathcal{A}, y \rangle_{y \in Y}$. The elements of I are indiscernible over Y if they are indiscernible in $\langle \mathcal{A}, y \rangle_{y \in Y}$.

Substructure Proviso. Suppose T is substructure complete, $\mathcal{A} \in \mathcal{K}(T)$ and $Y \subset A$. Let $\mathcal{B}$ be the least substructure of $\mathcal{A}$ whose universe contains Y. The elements of B are named by terms of the form $t(\underline{y}_1, \ldots, \underline{y}_n)$, where $t(x_1, \ldots, x_n)$ is a term of the language of T and $y_1, \ldots, y_n \in Y$. It will prove convenient to ignore the distinction between Y and $\mathcal{B}$. Thus "$p \in SY$" will mean "$p \in S\mathcal{B}$", and "$Y(a)$" will mean "$\mathcal{B}(a)$". Thus "$\langle a_1, \ldots, a_n \rangle$ realizes an atom over Y" in place of "$\langle a_1, \ldots, a_n \rangle$ realizes an atom over $\mathcal{B}$".

Theorem 35.4. *Let I be an infinite set of order indiscernibles over Y in $\mathcal{A}$. Suppose $p \in SY$ is the 1-type realized by all members of I in $\mathcal{A}$. If p has a (Morley) rank, then I is a set of indiscernibles over Y in $\mathcal{A}$.*

Proof. A small modification of the proof of 35.1. Choose $G(x_1, \ldots, x_n)$, a formula in the language of $T(\langle \mathcal{A}, y \rangle_{y \in Y})$, as in 35.1 so that

$$\mathcal{A} \models G(\underline{i}_1, \ldots, \underline{i}_n)$$
$$\mathcal{A} \models\sim G(\underline{i}_1, \ldots, \underline{i}_{k-1}, i_{k+1}, \underline{i}_k, \underline{i}_{k+2}, \ldots, \underline{i}_n)$$

whenever $i_1 < \cdots < i_n \in I$. Choose $\mathcal{B}$ as in 35.1 save for the additional requirement that every $r \in R$ realizes p. Then distinct members of R realize distinct 1-types over $Y \cup Q$, where Q is the set of rationals, and so p has uncountably many pre-images in $S(Y \cup Q)$. It is safe to assume Y is finite, since only finitely many members of Y are mentioned in $G(x_1, \ldots, x_n)$. But then (as in 31.6) p has only countably many pre-images in $S(Y \cup Q)$, since every pre-image of p has a rank and $S(Y \cup Q)$ has a countable base. $\square$

Suppose T is complete and substructure complete, $\mathcal{A} \in \mathcal{K}(T)$, $Y \subset A$ and $\{a_\delta \mid \delta < \alpha\} \subset A$. For each $\delta < \alpha$, let p_δ be the 1-type realized by a_δ over $Y \cup \{a_\gamma \mid \gamma < \delta\}$. Suppose p_0 has a (Morley) rank. $\{a_\delta \mid \delta < \alpha\}$ is a Morley sequence over Y in $\mathcal{A}$ if p_δ is a pre-image of p_γ of the same rank and degree as p_γ whenever $\gamma < \delta$. The rank and degree of $\{a_\delta \mid \delta < \alpha\}$ is that of p_0.

Lemma 35.5. *Every infinite Morley sequence over Y in $\mathcal{A}$ is a set of indiscernibles over Y in $\mathcal{A}$.*

Proof. By 35.4 it suffices to show

$$\mathcal{A} \models F(\underline{a}_{\delta_1}, \ldots, \underline{a}_{\delta_n}) \leftrightarrow F(\underline{a}_{\gamma_1}, \ldots, \underline{a}_{\gamma_n})$$

whenever $\delta_1 < \cdots < \delta_n$, $\gamma_1 < \cdots < \gamma_n$, and $F(x_1, \ldots, x_n)$ is a formula in the language of $T(\langle \mathcal{A}, y \rangle_{y \in Y})$. When $n = 1$, every a_δ realizes p_0. Fix $n > 1$. By induction on n there exists an isomorphism j such that

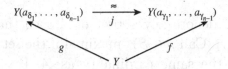

and $j a_{\delta_i} = a_{\gamma_i}$ for $1 \leq i \leq n$. Let p (respectively q) be the 1-type realized by a_{δ_n} (respectively a_{γ_n}) over $Y(a_{\delta_1}, \ldots, a_{\delta_{n-1}})$ (respectively $Y(a_{\gamma_1}, \ldots, a_{\gamma_{n-1}})$). It need only be shown that $(Sj)q = p$.

Clearly p is a pre-image of p_0, and p_{δ_n} is a pre-image of p. By the rank rule (29.3), rank $p_0 \geq$ rank $p \geq$ rank p_{δ_n}. Since $\{a_\delta \mid \delta < \alpha\}$ is a Morley sequence, rank $p_0 =$ rank p_{δ_n}, and consequently rank $p =$ rank p_0. A similar application of the degree rule (29.5) shows that deg $p =$ deg p_0, and that p is the *unique* pre-image of p_0 in $SY(a_{\delta_1}, \ldots, a_{\delta_{n-1}})$ of the same rank and degree as p_0. But the same reasoning fits q, so q is the unique pre-image of p_0 in $SY(a_{\gamma_1}, \ldots, a_{\gamma_n})$ of the same rank and degree as p_0. It follows that $(Sj)q = p$, since Sj, being a homeomorphism, preserves rank and degree, and since $(Sg)(Sj) = Sf$. $\square$

Proposition 35.6. *Every Morley sequence over Y in $\mathcal{A}$ of degree 1 is a set of indiscernibles over Y in $\mathcal{A}$.*

Proof. By 35.5 only finite Morley sequences need be considered. Let $\{a_i \mid i \leq n\}$ be a Morley sequence over Y of degree 1. Thus a_n realizes p_n over $Y(a_0, \ldots, a_{n-1})$ and deg $p_n = 1$. By the degree rule (29.5), p_n has a unique pre-image, call it p_{n+1}, in $SY(a_0, \ldots, a_n)$ of the same rank and degree as p_n. Choose a_{n+1} to realize p_{n+1}. Continue in this fashion until $\{a_i \mid i \leq n\}$ has been extended to a Morley sequence $\{a_i \mid i < \omega\}$ over Y. Then apply 35.5. (Note that it may be necessary to go beyond $\mathcal{A}$ to choose a_i $(i > n)$.) $\square$

Theorem 35.7 (M. Morley). *Suppose T is totally transcendental, $\mathcal{A} \in \mathcal{K}(T)$, $Z \subset A$, card Z < card A and card $\mathcal{A}$ is an uncountable regular cardinal. Then there exists a set I of indiscernibles over Z in $\mathcal{A}$ such that card $I =$ card $\mathcal{A}$.*

Proof. Let Y denote any subset of A such that $Z \subset Y$ and card Y < card A. Call $p \in SY$ maximal if the set of realizations of p in $\mathcal{A}$ has the same cardinality as $\mathcal{A}$. By 31.6 and 19.1, card SY < card A. Since card A is regular, there must exist a maximal $p \in SY$. For each Y choose a maximal $pY \in SY$ of the least possible rank and of the least possible degree in that rank. Then choose Y^0 so that pY^0 has the least possible rank and the least possible degree in that rank.

Suppose $Y \supset Y^0$. Clearly pY^0 has a maximal pre-image in SY; let q be such a pre-image. Then rank $q \geq$ rank pY^0 by definition of Y^0, and so rank $q =$ rank pY^0 by the rank rule. Similarly deg $q =$ deg pY^0 by the degree rule. Thus maximal pre-images of pY^0 have the same rank and degree as pY^0. It follows from the degree rule that pY^0 has a unique maximal pre-image in SY.

A Morley sequence $\{a_\delta \mid \delta <$ card $\mathcal{A}\}$ is defined by induction on δ.

1. $p_0 = pY^0$.
2. a_δ realizes p_δ.
3. Assume $p_\delta \in S(Y^0 \cup \{a_\gamma \mid \gamma < \delta\})$ is a maximal pre-image of p_0. Let $p_{\delta+1} \in S(Y^0 \cup \{a_\gamma \mid \gamma \leq \delta\})$ be a maximal pre-image of p_δ.
4. Let $p_\lambda = \cup \{p_\delta \mid \delta < \lambda\}$. Assume p_δ is a maximal pre-image of p_0 for every $\delta < \lambda$. Suppose $b \in A$ realizes p_0 but not p_δ. Then b realizes some nonmaximal $q \in S(Y \cup \{a_\gamma \mid \gamma < \delta\})$, since p_δ

is the unique maximal pre-image of p in $S(Y \cup \{a_\gamma \mid \gamma < \delta\})$. Consequently p_λ is a maximal pre-image of p_0.

Let $I = \{a_\delta \mid \delta < \text{card } \mathcal{A}\}$. I is indiscernible over Y in $\mathcal{A}$ by 35.5. $\qquad \square$

Lemma 35.8 (S. Shelah). *Suppose T is totally transcendental, $\mathcal{A} \subset \mathcal{B} \subset \mathcal{C} \in \mathcal{K}(T)$, $\mathcal{B}$ is finitely generated over $\mathcal{A}$, and I is a set of indiscernibles over $\mathcal{A}$ in $\mathcal{C}$. Then there exists a finite $J \subset I$ such that $I - J$ is indiscernible over $\mathcal{B}(J)$ in $\mathcal{C}$.*

Proof. Assume $\mathcal{B} = \mathcal{A}(b)$. Choose a finite $J \subset I$ so that the rank and degree of the type p_J realized by b over $\mathcal{A}(J)$ equal the rank and degree of the type p_I realized by b over $\mathcal{A}(I)$. J exists by 29.2, since $\mathcal{A}(I)$ is the direct limit of those substructures of $\mathcal{A}(I)$ that are finitely generated over $\mathcal{A}$.

Let I_1 and I_2 be finite subsets of $I - J$ of the same cardinality. Let f map I_1 one-one onto I_2. Since I is indiscernible over $\mathcal{A}$, f can be extended to an isomorphism

$$f \colon \mathcal{A}(J)(I_1) \xrightarrow{\approx} \mathcal{A}(J)(I_2)$$

such that $f \mid A(J) = 1_{A(J)}$. Let p_1 (respectively p_2) be the type realized by b over $\mathcal{A}(J)(I_1)$ (respectively $\mathcal{A}(J)(I_2)$). It suffices to show $Sfp_2 = p_1$. p_2 and p_1 are images of p_I and pre-images of p_J. It follows from the definition of p_J (and the rank and degree rules) that p_1 and p_2 have the same rank and degree as p_J. Thus p_1 is the unique pre-image of p_J in $S\mathcal{A}(J)(I_1)$ of the same rank and degree as p_J. But Sfp_2 is also a pre-image of p_J of the same rank and degree as p_J, since Sf is a homeomorphism:

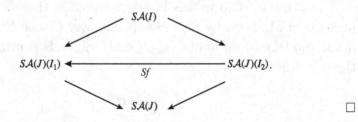

$\qquad \square$

Lemma 35.9 (S. Shelah). *Suppose T is totally transcendental, $\mathcal{A} \in \mathcal{K}(T)$, and $\mathcal{B}$ is a prime model extension of $\mathcal{A}$. Then every set of indiscernibles over $\mathcal{A}$ in $\mathcal{B}$ is countable.*

Proof. Suppose I is an infinite set of indiscernibles over $\mathcal{A}$ in $\mathcal{B}$. Let $F(x)$ be a formula in the language of $T \cup D\mathcal{B}$. By 35.8 there is a finite $J \subset I$ such that either $\mathcal{B} \models F(\underline{i})$ for all $i \in I - J$ or $\mathcal{B} \models {\sim}F(\underline{i})$ for all $i \in I - J$. Define

$$p^I = \{F(x) \,|\, \mathcal{B} \models F(\underline{i}) \quad \text{for all but finitely many } i \in I\}.$$

By 35.8 $p^I \in S\mathcal{B}$. (p^I is the "average" type of I in $\mathcal{B}$.)

In the hope of reaching a contradiction, choose I, an uncountable set of indiscernibles over $\mathcal{B}$ in $\mathcal{A}$, so that p^I has the least possible rank and the least possible degree in that rank. By 29.2 there is a $\mathcal{C} \subset \mathcal{B}$, finitely generated over $\mathcal{A}$, such that $p^I_{\mathcal{C}}$, the projection of p^I in $S\mathcal{C}$, has the same rank and degree as p^I. Then for every $\mathcal{D}$ such that $\mathcal{C} \subset \mathcal{D} \subset \mathcal{B}$, $p^I_{\mathcal{D}}$ has the same rank (say α) and degree (say m) as p^I.

By 35.8 there is a finite $J \subset I$ such that $K = I - J$ is indiscernible over $\mathcal{C}$ in $\mathcal{B}$. Sequences $\{p_n \,|\, n < \omega\}$ and $\{i_n \,|\, n < \omega\}$ are defined by induction.

1. $p_0 \in S\mathcal{C}$, and every $i \in K$ realizes p_0.
2. $i_n \in K$ and i_n realizes p_n.
3. By 35.8 all but finitely many $i \in K$ realize the same type over $\mathcal{C}(i_0, \ldots, i_n)$; let p_{n+1} be that type. Observe that $p_n = p^I_{\mathcal{C}}(i_0, \ldots, i_n)$ and so has rank α and degree m thanks to the choice of $\mathcal{C}$.

The proof of 32.3 makes it safe to assume that $\mathcal{B}$ is Morley prime over $\mathcal{A}$. Then by 32.8 $\mathcal{B}$ is prime over $\mathcal{C}$. Let $\mathcal{B}^* \subset \mathcal{B}$ be a prime model extension of $\mathcal{C}(i_n \,|\, n < \omega)$. Since $\mathcal{B}$ is prime over $\mathcal{C}$, there is a map

$$f \colon \mathcal{B} \overset{\equiv}{\to} \mathcal{B}^*$$

such that $f \mid C = 1_C$. Let $K^* = f[K]$. Then K^* is indiscernible over C in B^*. Let p_n^* be the type realized by all but finitely many $i \in K^*$ over $C(i_0, \ldots, i_{n-1})$. Clearly $p_0 = p_0^*$. An induction on n shows $p_n = p_n^*$. Suppose $p_n = p_n^*$. Then rank $p_n^* = $ rank $p_n = $ rank $p_C^I = \alpha$. The rank rule yields rank $p_{n+1}^* \leq \alpha$. Let p^{K^*} be the "average" type of K^* in B. K^* is indiscernible over A in B, so rank $p^{K^*} \geq$ rank $p^I = \alpha$ thanks to the choice of I. But rank $p_{n+1}^* = $ rank $p_C^{K^*}(i_0, \ldots, i_n) \geq$ rank p^{K^*}. Thus rank $p_{n+1}^* = $ rank $p_{n+1} = \alpha$. But then the degree rule implies $p_{n+1} = p_{n+1}^*$, since p_{n+1} and p_{n+1}^* are pre-images of p_n of the same rank as p_n, and since deg $p_{n+1} = $ deg $p_n = m$.

Let $p_\omega = \cup \{p_n \mid n < \omega\} = \cup \{p_n^* \mid n < \omega\}$. Since K^* is uncountable, 35.8 implies there is an $i^* \in K^*$ that realizes p_ω. By 32.6 p_ω is an isolated point of $SC(i_n \mid n < \omega)$. Let $F(x, \underline{i}_0, \ldots, \underline{i}_n) \in p_{n+1}$ generate p_ω. Then

$$T \cup D(C(i_n \mid n < \omega)) \vdash F(x, \underline{i}_0, \ldots, \underline{i}_n) \to x \neq \underline{i}_{n+1},$$

since $i^* \neq i_{n+1}$. This last is nonsense because i^* and i_{n+1} are indiscernible over $C(i_0, \ldots, i_n)$. $\qquad\square$

Theorem 35.10 (S. Shelah). *If T is totally transcendental and $\kappa \geq \omega$, then T has a saturated model of cardinality κ.*

Proof. By 19.4 and 31.6 T has a special model A if cardinality κ. Thus $A = \cup \{A_\gamma \mid \gamma < \alpha\}$ where $\{A_\gamma \mid \gamma < \alpha\}$ is an elementary chain of saturated models. Suppose $Y \subset A$ is such that card $Y < \kappa$. Fix $p \in SY$ with the intent of showing p is realized in A. Let $q \in SA$ be a pre-image of p. It follows from 29.2 that there is a finitely generated $C \subset A$ such that q_C, the projection of q in SC, has the same rank and degree as q. Then for every D such that $C \subset D \subset A$, q_D has the same rank and degree as q.

Assume α is a limit ordinal; if not, A is saturated. Thus $C \subset A_\gamma$ for some $\gamma < \alpha$. Assume $C \subset A_0$ and card $A_{\gamma 1} < $ card $A_{\gamma 2}$

when $\gamma 1 < \gamma 2 < \alpha$. A Morley sequence $\{i_\delta \,|\, \delta < \kappa\} \subset A$ is defined by induction on $\delta < \kappa$.

1. Let $q_\delta \in SC(i_\rho \,|\, \rho < \delta)$ be the projection of q. Then q_δ has the same rank and degree as q.
2. Assume there is a $\gamma < \alpha$ such that $C(i_\rho \,|\, \rho < \delta) \subset A_\gamma$ and card $A_\gamma > \max(\omega, \text{card } \delta)$; let $\gamma\delta$ be the least such γ. Choose $i_\delta \in A_{\gamma\delta}$ to relaize q_δ; such an i_δ exists since $A_{\gamma\delta}$ is saturated.

   Let $B \supset A$ be κ^+-saturated in order to extend $\{i_\delta \,|\, \delta < \kappa\}$ to a Morley sequence $\{i_\delta \,|\, \delta < 2\kappa\} \subset B$.
3. Let $q_\kappa = q \in SA$. By 31.8 deg $q_\kappa = 1$.
4. Assume $\kappa \leq \delta < 2\kappa$. Choose $i_\delta \in B$ to realize q_δ. Assume deg $q_\delta = 1$. Let $q_{\delta+1} \in SA(i_\rho \,|\, \rho \leq \delta)$ be the unique pre-image of q_δ of the same rank and degree as q_δ. $q_{\delta+1}$ exists by the degree rule, since deg $q_\delta = 1$.
5. Let $q_\lambda = \cup \{q_\delta \,|\, \delta < \lambda\}$ when λ is a limit ordinal.

To check that $\{i_\delta \,|\, \delta < 2\kappa\}$ is a Morley sequence, define $r_\delta \in SC(i_\rho \,|\, \rho < \delta)$ to be the type realized by i_δ. If $\delta < \kappa$, then $r_\delta = q_\delta$. If $\kappa \leq \delta < 2\kappa$, then $r_{\delta+1}$ is a pre-image of r_δ, since $q_{\delta+1}$ is a pre-image of $r_{\delta+1}$ and also of r_δ. It is immediate by induction on δ that r_δ has the same rank and degree as r_0 for all $\delta < 2\kappa$. By 35.5 $I = \{i_\delta \,|\, \delta < 2\kappa\}$ is indiscernible over C in B. It follows from 35.8 that there exists a $J \subset I$ such that card $J < \kappa$ and $I - J$ is indiscernible over Y in B. Choose $i^* \in \{i_\delta \,|\, \kappa \leq \delta < 2\kappa\} - J$. Then i^* realizes p in B, since q_κ is a pre-image of p. Choose $i \in \{i_\delta \,|\, \delta < \kappa\} - J$. Then i realizes p in A, since i and i^* are indiscernible over Y. $\qquad \square$

Exercise 35.11 (S. Shelah). Suppose T is totally transcendental and $\kappa \geq \omega$. Show there exists a model A of T and a set $I \subset A$ such that card $I = \text{card } A = \kappa$ and every one-one map of I into I can be extended to an automorphism of A.

Section 36

Shelah's Uniqueness Theorem

Suppose T is totally transcendental, $\mathcal{A} \in \mathcal{K}(T)$, and $\mathcal{B}_1$ and $\mathcal{B}_2$ are prime model extensions of $\mathcal{A}$. Then $\mathcal{B}_1$ and $\mathcal{B}_2$ are isomorphic over $\mathcal{A}$ according to Theorem 36.2. The isomorphism between $\mathcal{B}_1$ and $\mathcal{B}_2$ owes its existence principally to Lemma 35.9, which says no prime model extension of $\mathcal{A}$ contains an uncountable set of indiscernibles over $\mathcal{A}$. The definition of the isomorphism is by induction on a slight modification of Morley rank.

Let F be a closed subset of $S\mathcal{A}$. The ordinal rank of F is the least $\gamma \geq$ rank p for every $p \in F$. If the ordinal rank of $F = \alpha$, then the compactness of $S\mathcal{A}$ implies that at least one, but not infinitely many, $p \in F$ are of rank α.

Let n be the highest degree attained among those $p \in F$ of rank α, and let d be the number of $p \in F$ of rank α and degree n. The rank of F is (α, n, d). (α, n_1, d_1) is less than (α_2, n_2, d_2) if $\alpha_1 < \alpha_2$; or if $\alpha_1 = \alpha_2$ and $n_1 < n_2$; or if $\alpha_1 = \alpha_2$, $n_1 = n_2$ and $d_1 < d_2$. If $F(x)$ is a formula in the language of $T \cup D\mathcal{A}$, then $\{p \mid p \in S\mathcal{A} \ \& \ F(x) \in p\}$ is a closed set of $S\mathcal{A}$. The A-rank of $F(x)$ is the rank of $\{p \mid p \in S\mathcal{A} \ \& \ F(x) \in p\}$.

Theorem 36.1 (S. Shelah). *Suppose T is totally transcendental, $\mathcal{A}$ (respectively $\mathcal{A}^*$) $\in \mathcal{K}(T)$, $\mathcal{B}$ (respectively $\mathcal{B}^*$) is an atomic model extension of $\mathcal{A}$ (respectively $\mathcal{A}^*$), and $f : A \to A^*$*

is an onto map such that

$$\langle \mathcal{B}, a \rangle_{a \in A} \equiv \langle \mathcal{B}^*, fa \rangle_{a \in A}.$$

Suppose further that every set of indiscernibles over $\mathcal{A}$ (respectively $\mathcal{A}^*$) in $\mathcal{B}$ (respectively $\mathcal{B}^*$) is countable. Let $F(x)$ be a formula in the language of $T \cup D\mathcal{A}$; define

$$C = A \cup \{b \,|\, b \in B \;\&\; \mathcal{B} \models F(\underline{b})\},$$

$$C^* = A^* \cup \{b \,|\, b \in B^* \;\&\; \mathcal{B}^* \models F(\underline{b})\}.$$

Then f can be extended to an onto map $g\colon C \to C^*$ such that

$$\langle \mathcal{B}, c \rangle_{c \in C} \equiv \langle \mathcal{B}^*, gc \rangle_{c \in C}.$$

Proof. By induction on the A-rank of $F(x)$. First Suppose that rank is (α, n, d), where $d > 1$. Choose $p \in S\mathcal{A}$ so that $F(x) \in p$, $\operatorname{rank} p = \alpha$ and $\deg p = n$. Choose $G(x)$ so that (i) $G(x) \in p$, and (ii) $G(x) \in q$ implies $q = p$ or $\operatorname{rank} q < \alpha$. Let

$$D = A \cup \{b \,|\, b \in B \;\&\; \mathcal{B} \models F(\underline{b}) \;\&\; \sim G(\underline{b})\},$$

$$D^* = A \cup \{b \,|\, b \in B^* \;\&\; \mathcal{B}^* \models F(\underline{b}) \;\&\; \sim G(\underline{b})\}.$$

Then f can be extended to an onto map $h\colon D \to D^*$ such that

$$\langle \mathcal{B}, d \rangle_{d \in D} \equiv \langle \mathcal{B}^*, hd \rangle_{d \in D},$$

since the A-rank of $F(x) \;\&\; \sim G(x)$ is $(\alpha, n, d-1)$. It follows from the rank and degree rules of Sec. 29 that the D-rank of $F(x) \;\&\; G(x)$ is less than the A-rank of $F(x)$. So h can be extended to an onto map $g\colon C \to C^*$ such that

$$\langle \mathcal{B}, c \rangle_{c \in C} \equiv \langle \mathcal{B}^*, gc \rangle_{c \in C}.$$

Now suppose the A-rank of $F(x)$ is $(\alpha, n, 1)$. Thus there is a unique $p \in S\mathcal{A}$ such that: $F(x) \in p$, $\operatorname{rank} p = \alpha$ and $\deg p = n$. Sequences $\{p_\delta \,|\, \delta \le \delta_0\}$ and $\{b_\delta \,|\, \delta < \gamma_0\}$ are defined by induction.

1. $p_0 = p$.
2. Choose $b_\delta \in C - A$ to realize p_δ. If that is not possible, then $\gamma_0 = \delta = \delta_0$.
3. Assume $p_\delta \in S\mathcal{A}(b_\gamma \,|\, \gamma < \delta)$. $p_{\delta+1} \in S\mathcal{A}(b_\gamma \,|\, \gamma \leq \delta)$ must be a pre-image of p_δ of the same rank and degree as p_δ. If that is not possible, then $\delta_0 = \delta$ and $\gamma_0 = \delta + 1$.
4. $p_\lambda = \cup\{p_\delta \,|\, \delta < \lambda\}$.

If $\delta_0 \geq \omega$, then $\{b_\delta \,|\, \delta < \gamma_0\}$ is a set of indiscernibles over $\mathcal{A}$ in $\mathcal{B}$ by 35.5. Hence γ_0 is countable. If $b \in C - A$ and realizes p, then the maximality of $\{b_\delta \,|\, \delta < \gamma_0\}$ compels b to realize a 1-type of $S\mathcal{A}(b_\delta \,|\, \delta < \gamma_0)$ of lower rank or degree than p. Thus $\{b_\delta \,|\, \delta < \gamma_0\}$ is a countable "transcendence base" for the realizations of p in $B - A$. Reorder the b_δ's so that $\{b_\delta \,|\, \delta < \gamma_0\} = \{b_m \,|\, m < \alpha_0\}$ for some $\alpha_0 \leq \omega$. α_0 is the "dimension" of p in $\mathcal{B}$. Observe that every $b \in C$ realizes a 1-type of $S\mathcal{A}(b_m \,|\, m < \alpha_0)$ of rank $< \alpha$ or of rank α and degree $< n$.

Let $p^* = (Sf)^{-1}p \in S\mathcal{A}^*$. Repeat the above construction with $\mathcal{A}^*$, $\mathcal{B}^*$, C^* and p^* in place of $\mathcal{A}$, $\mathcal{B}$, C and p to obtain $\{b_m^* \,|\, m < \alpha_0^*\}$. If $j < \omega$ and $\alpha_0 \geq j$, then $\alpha_0^* \geq j$, since $\mathcal{B}$ is atomic over $\mathcal{A}$ and $\langle \mathcal{B}, a \rangle_{a \in A} \equiv \langle \mathcal{B}^*, fa \rangle_{a \in A}$. On similar grounds $\alpha_0^* \geq j$ implies $\alpha_0 \geq j$. Consequently $\alpha_0 = \alpha_0^*$.

$g = \cup\{g_m \,|\, m < \omega\}$ is defined by induction on m. Let $g_0 = f$. Fix $m \geq 0$. Assume g_m has been defined so that domain $g_m \subset C$, range $g_m \subset C^*$, $\mathcal{B}$ (respectively $\mathcal{B}^*$) is atomic over domain g_m (respectively range g_m), and

$$\langle \mathcal{B}, c \rangle_{c \in \text{ domain } g_m} \equiv \langle \mathcal{B}^*, g_m c \rangle_{c \in \text{ domain } g_m}.$$

Case 0. $m + 1 \equiv 0(3)$. Let $k = \frac{m+1}{3}$. The purpose of case 0 is to put b_k into domain g_{m+1}. If $k \geq \alpha_0$ or $b_k \in$ domain g_m, then $g_{m+1} = g_m$. Suppose k is otherwise. The assumptions on g_m yield a $c^* \in C^*$ such that

$$\langle \mathcal{B}, c, b_k \rangle_{c \in \text{ domain } g_m} \equiv \langle \mathcal{B}^*, g_m c, c^* \rangle_{c \in \text{ domain } g_m}.$$

Extend g_m to g_{m+1} so that $g_{m+1}b_k = c^*$. By 32.7 $\mathcal{B}$ (respectively $\mathcal{B}^*$) is atomic over domain g_{m+1} (respectively range g_{m+1}).

Case 1. $m + 1 \equiv 2$ (3). Let $k = \frac{m}{3}$. Similar is case 0 with b_k and domain g_{m+1} replaced by b_k^* and range g_{m+1}.

Case 2. $m + 1 \equiv 2$ (3). The object of case 2 is to extend g_m to g_{m+1} so that domain g_{m+1} (respectively range g_{m+1}) contains every $b \in C$ (respectively C^*) with the following property: b realizes a 1-type $q \in S(\text{domain } g_m)$ (respectively $S(\text{range } g_m)$) of lower rank or degree than that of p. (Recall that every $b \in C$ realizes a 1-type of $S\mathcal{A}(b_m \,|\, m < \alpha_0)$ of lower rank or degree than that of p.) Let $\{q_\gamma \,|\, \gamma < \rho_0\}$ be an enumeration of all $q \in S(\text{domain } g_m)$ of lower rank or degree than that of p such that q is satisfied by some $b \in C$. Since $\mathcal{B}$ is atomic over domain g_m, every q_γ is an isolated point of $S(\text{domain } g_m)$. Let $F_\gamma(x)$ be a formula in the language of $T \cup D(\text{domain } g_m)$ such that q_γ is generated by $F_\gamma(x)$. The (domain g_m)-rank of $F_\gamma(x)$ is less than the A-rank of $F(x)$, since rank $q_\gamma <$ rank p.

Let $q_\gamma^* = (Sg_m)^{-1}q_\gamma \in S(\text{range } g_m)$. Then $\{q_\gamma^* \,|\, \gamma < \rho_0\}$ is an enumeration of all $q \in S(\text{range } g_m)$ of lower rank or degree than that of p such that q is satisfied by some $b \in C^*$. Define $F_\gamma^*(x)$ in a similar fashion.

$g_{m+1} = \cup \{g_{m+1}^\gamma \,|\, \gamma < \rho_0\}$ is defined by induction.

1. $g_{m+1}^0 = g_m$.
2. Assume that g_{m+1}^γ has been defined so that $\mathcal{B}$ (respectively $\mathcal{B}^*$) is atomic over domain g_{m+1}^γ (respectively range g_{m+1}^γ) and

$$\langle \mathcal{B}, b \rangle_{b \in \text{ domain } g_{m+1}^\gamma} \equiv \langle \mathcal{B}^*, g_{m+1}^\gamma b \rangle_{b \in \text{ domain } g_{m+1}^\gamma}.$$

Clearly the (domain g_{m+1}^γ)-rank of $F_\gamma(x)$ is less than the A-rank of $F(x)$. Consequently g_{m+1}^γ can be extended to $g_{m+1}^{\gamma+1}$

so that

$$\text{domain } g_{m+1}^{\gamma+1} = \text{ domain } g_{m+1}^{\gamma} \cup \{b \mid \mathcal{B} \models F_{\gamma}(\underline{b})\},$$

$$\text{range } g_{m+1}^{\gamma+1} = \text{ range } g_{m+1}^{\gamma} \cup \{b \mid \mathcal{B}^* \models F_{\gamma}^*(b)\},$$

$$\langle \mathcal{B}, c \rangle_{c \in \text{ domain } g_{m+1}^{\gamma+1}} \equiv \langle \mathcal{B}^*, g_{m+1}^{\gamma+1} c \rangle_{c \in \text{ domain } g_{m+1}^{\gamma+1}}.$$

By 32.9 $\mathcal{B}$ is atomic over domain $g_{m+1}^{\gamma+1}$, since domain $g_{m+1}^{\gamma+1}$ is a normal extension of domain g_{m+1}^{γ}. By the same reasoning $\mathcal{B}^*$ is atomic over range $g_{m+1}^{\gamma+1}$.

3. $g_{m+1}^{\lambda} = \cup \{g_{m+1} \mid \gamma < \lambda\}$. Domain g_{m+1}^{γ} is a normal extension of domain g_{m+1} in $\mathcal{B}$, so 32.9 implies $\mathcal{B}$ is atomic over domain g_{m+1}^{γ}. $\qquad\square$

Corollary 36.2 (S. Shelah). *If T is totally transcendental and $\mathcal{A} \in \mathcal{K}(T)$, then any two prime model extensions of $\mathcal{A}$ are isomorphic over $\mathcal{A}$.*

Proof. Let $\mathcal{B}$ and $\mathcal{B}^*$ be prime model extensions of $\mathcal{A}$. By 35.9 every set of indiscernibles over $\mathcal{A}$ in $\mathcal{B}$ (respectively $\mathcal{B}^*$) is countable. By 32.2 and 32.6, $\mathcal{B}$ (respectively $\mathcal{B}^*$) is atomic over $\mathcal{A}$. It follows from 36.1 that the identity map on A can be extended to an isomorphism between $\mathcal{B}$ and $\mathcal{B}^*$: let $F(x)$ be $x = x$. $\qquad\square$

The conclusion of 36.2 stands when the hypothesis of total transcendentality is weakened to quasi-total transcendentality (Exercise 36.6). Does it persist when the hypothesis is weakened still further to read: the isolated points of $S\mathcal{A}$ are dense in $S\mathcal{A}$ for every $\mathcal{A} \in \mathcal{K}(T)$?

Suppose T is totally transcendental and $\mathcal{A} \in \mathcal{K}(T)$. Theorem 32.3 affords a prime model extension of $\mathcal{A}$. Corollary 36.2 permits reference to *the* prime model extension of $\mathcal{A}$. 32.3 and 36.2 will be applied in Sec. 41 to define the differential closure of a differential field of characteristic 0.

Corollary 36.3. *Suppose T is totally transcendental, $\mathcal{A} \in \mathcal{K}(T)$, and $\mathcal{C}$ is a prime model extension of $\mathcal{A}$. If $\mathcal{A} \subset \mathcal{B} \subset \mathcal{C}$ and $\mathcal{C}$ is atomic over $\mathcal{B}$, then every automorphism of $\mathcal{B}$ can be extended to an automorphism of $\mathcal{C}$.*

Corollary 36.4 (S. Shelah). *Suppose T is totally transcendental, $\mathcal{A} \in \mathcal{K}(T)$, and $\mathcal{B}$ is a model extension of $\mathcal{A}$. Then (i) is equivalent to (ii).*

(i) *$\mathcal{B}$ is prime over $\mathcal{A}$.*
(ii) *$\mathcal{B}$ is atomic over $\mathcal{A}$, and every set of indiscernibles over $\mathcal{A}$ in $\mathcal{B}$ is countable.*

Proof. (i) implies (ii) by 32.2, 32.6, and 35.9. (ii) implies (i) by 36.1.   □

Corollary 36.5 (S. Shelah). *Suppose T is totally transcendental, $\mathcal{A} \in \mathcal{K}(T)$, and $\mathcal{B}$ is a model extension of $\mathcal{A}$. Then (i) is equivalent to (ii).*

(i) *$\mathcal{B}$ is minimal over $\mathcal{A}$.*
(ii) *$\mathcal{B}$ is atomic over $\mathcal{A}$, and every set of indiscernibles over $\mathcal{A}$ in $\mathcal{B}$ is finite.*

Proof. Suppose (i) holds. Then $\mathcal{B}$ is prime over $\mathcal{A}$ by 32.4, hence atomic over $\mathcal{A}$ by 32.6. Suppose I were an infinite set of indiscernibles over $\mathcal{A}$ in $\mathcal{B}$. Fix $i \in I$. Let $\mathcal{C} \subset \mathcal{B}$ be a prime model extension of $\mathcal{A}(I - \{i\})$. Then $i \in C$, since $\mathcal{B}$ is minimal over $\mathcal{A}$. In addition i realizes some isolated $p \in S\mathcal{A}(I - \{i\})$, since $\mathcal{C}$ is atomic over $\mathcal{A}(I - \{i\})$. Let $F(x)$ generate p, and let J be the finite subset of $I - \{i\}$ mentioned in $F(x)$. Choose $j \in (I - \{i\}) - J$. Then

$$T \cup D\mathcal{A}(I - \{i\}) \vdash F(x) \rightarrow x \neq \underline{j}$$
$$T \cup D\mathcal{A}(I - \{i\}) \vdash\, \sim F(\underline{j}).$$

But $\mathcal{B} \models F(\underline{j})$ is a consequence of $\mathcal{B} \models F(\underline{i})$, because j and i are indiscernible over $\mathcal{A}(J)$.

Now suppose (ii) holds and (i) fails. By 36.4 $\mathcal{B}$ is prime over $\mathcal{A}$. It follows there is a $\mathcal{B}^* \succ \mathcal{B}$ such that $\mathcal{B}^* \neq \mathcal{B}$ and $\mathcal{B}^*$ and $\mathcal{B}$ are isomorphic over $\mathcal{A}$. An elementary chain $\{\mathcal{B}_\delta \mid \delta < (\text{card } \mathcal{A})^+\}$ is defined by induction with the intent of locating an infinite set of indiscernibles over $\mathcal{A}$ in $\mathcal{B}$.

1. $\mathcal{B}_0 = \mathcal{B}$.
2. Assume $\mathcal{B}_\delta$ is isomorphic to $\mathcal{B}$ over $\mathcal{A}$. Let $\mathcal{B}_{\delta+1}$ bear the same relation to $\mathcal{B}_\delta$ that $\mathcal{B}^*$ bears to $\mathcal{B}$. Thus $\mathcal{B}_{\delta+1} \succ \mathcal{B}_\delta$, $\mathcal{B}_{\delta+1} \neq \mathcal{B}_\delta$, and $\mathcal{B}_{\delta+1}$ and $\mathcal{B}_\delta$ are isomorphic over $\mathcal{A}$.
3. Let $\mathcal{B}_\lambda = \cup\{\mathcal{B}_\delta \mid \delta < \lambda\}$. Assume $\mathcal{B}_\delta$ is isomorphic to $\mathcal{B}$ over $\mathcal{A}$ for every $\delta < \lambda$. Then $\mathcal{B}_\lambda$ is atomic over $\mathcal{A}$. If $\mathcal{B}_\lambda$ has an uncountable set of indiscernibles over $\mathcal{A}$, then for some $\delta < \lambda$: $\mathcal{B}_\delta$, hence $\mathcal{B}$, has an infinite set of indiscernibles over $\mathcal{A}$. If every set of indiscernibles over $\mathcal{A}$ in $\mathcal{B}_\lambda$ is countable, then $\mathcal{B}_\lambda$ is isomorphic to $\mathcal{B}$ over $\mathcal{A}$ by 36.4.

Let $\mathcal{C} = \cup\{\mathcal{B}_\delta \mid \delta < (\text{card } \mathcal{A})^+\}$. By 35.7 there is an uncountable set of indiscernibles over $\mathcal{A}$ in $\mathcal{C}$. But then for some $\delta < (\text{ card } \mathcal{A})^+$, $\mathcal{B}_\delta$, hence $\mathcal{B}$, has an infinite set of indiscernibles over $\mathcal{A}$. $\qquad\square$

Exercise 36.6. Suppose T is quasi-totally transcendental and $\mathcal{A} \in \mathcal{K}(T)$. Show any two prime model extensions of $\mathcal{A}$ are isomorphic over $\mathcal{A}$.

Section 37

Categoricity in Some Uncountable Power

Morley's categoricity theorem (37.4) follows from a type omitting lemma (37.2) based on a downward going Skolem–Löwenheim result (37.1).

Proposition 37.1 (M. Morley). *Suppose T is totally transcendental and has an uncountable unsaturated model. Then there exist a countable model A of T, $Y \subset A$, $p \in SY$ and $I \subset A$ such that p is not realized in A and I is an infinite set of indiscernibles over Y in A.*

Proof. Let B be an uncountable unsaturated model of T. Choose $Z \subset B$ and $q \in SZ$ so that card $Z <$ card B and q is not realized in B. It follows from 35.7 that there exists a countably infinite set I of indiscernibles over Z in B.

An expanding sequence $\{A_n \mid n < \omega\}$ of countable substructures of B is defined by induction.

1. A_0 is the least substructure of B whose universe contains I.
2. Assume $A_{2n} \subset B$ is countable. Clearly q is not realized in A_{2n}. So for each $a \in A_{2n}$ there must be a formula $F_a(x) \in q$ such that $B \models {\sim} F_a(\underline{a})$. Consequently A_{2n+1} can be chosen so that $B \supset A_{2n+1} \supset A_{2n}$, A_{2n+1} is countable, and no $a \in A_{2n}$ realizes $q_{2n+1} \in S(Z \cap A_{2n+1})$, where q_{2n+1} is the projection

160

of $q \in SZ$. For each $a \in A_{2n}$, those members of Z mentioned in $F_a(x)$ belong to $Z \cap A_{2n+1}$.

3. Assume $\mathcal{A}_{2n+1} \subset \mathcal{B}$ is countable. By 11.2 $\mathcal{A}_{2n+2}$ can be chosen so that $\mathcal{B} \succ \mathcal{A}_{2n+2} \supset \mathcal{A}_{2n+1}$ and $\mathcal{A}_{2n+2}$ is countable.

Let $\mathcal{A} = \cup \{\mathcal{A}_n \,|\, n < \omega\}$. 6.1 and 10.3 imply $\mathcal{A} \prec \mathcal{B}$. Let $Y = Z \cap A$, and $p \in SY$ be the projection of q. If p were realized in $\mathcal{A}$ by $a \in A_{2n}$, then q_{2n+1} would be realized in $\mathcal{A}_{2n}$. If I were not indiscernible over Y in $\mathcal{A}$, then I would not be indiscernible over Z in $\mathcal{B}$. $\qquad\square$

Lemma 37.2 (M. Morley). *Suppose T is totally transcendental, $\mathcal{A} \models T$, $Y \subset A$, $p \in SY$, p is not realized in $\mathcal{A}$, and I is an infinite set of indiscernibles over Y in $\mathcal{A}$. If $\kappa \geq \text{card } \mathcal{A}$, then T has a model $\mathcal{C}$ of cardinality κ such that $Y \subset C$ and p is not realized in $\mathcal{C}$.*

Proof. The compactness theorem furnishes a $\mathcal{B}$ and a J such that $\mathcal{B} \models T$, $Y \subset B$, $I \subset J \subset B$, J is indiscernible over Y in $\mathcal{B}$, and card $J = \kappa$. Let $\mathcal{C}$ be the prime model extension of $Y \cup J$ afforded by 32.4. Suppose c realizes p in $\mathcal{C}$. By 32.6 c is atomic over $Y \cup J$, hence atomic over $Y \cup J^*$ for some finite $J^* \subset J$. It follows from 32.1 (i) and 32.2 that p is realized in every model extension of $Y \cup J^*$, hence in every model extension of $Y \cup I$, and hence in $\mathcal{A}$. $\qquad\square$

The notion of Morleyization makes clear the harmless nature of the assumption of substructure completeness. The Morleyization $T^{\mathcal{M}}$ of a theory T is obtained by adding new relation symbols and axioms as follows. For each formula $F(x_1, \ldots, x_n)$ in the language of T, add the relation symbol R_F and the axiom

$$(x_1) \cdots (x_n)[R_F(x_1, \ldots, x_n) \leftrightarrow F(x_1, \ldots, x_n)].$$

Proposition 37.3. *Each formula in the language of* $T^{\mathcal{M}}$ *is provably equivalent in* $T^{\mathcal{M}}$ *to some formula in the language of* T. $T^{\mathcal{M}}$ *is substructure complete.* T *is complete iff* $T^{\mathcal{M}}$ *is complete.* T *is* ω-*stable iff* $T^{\mathcal{M}}$ *is totally transcendental.*

Proof. By induction on the number of steps needed to generate the formula of $T^{\mathcal{M}}$ in question. $\qquad\qquad\square$

T and $T^{\mathcal{M}}$ have essentially the same models. Each model of T ($T^{\mathcal{M}}$) has a unique expansion (reduction) to a model of $T^{\mathcal{M}}$ (T) obtained by adding (deleting) the appropriate relations. But T and $T^{\mathcal{M}}$ embody radically different conceptions of the notion of substructure of a model.

Theorem 37.4 (M. Morley). *Let* T *be a countable, complete theory satisfying* (i) *or* (ii) *for some uncountable cardinal* κ.

(i) T *is* κ-*categorical.*
(ii) *Every model of* T *of cardinal* κ *is* ω_1-*saturated.*

Then every uncountable model of T *is saturated, and consequently* T *is categorical in every uncountable power.*

Proof. 37.3 allows T to be substructure complete. By 34.4, 34.6 and 31.6, T is totally transcendental. If T had an uncountable unsaturated model, then 37.1 and 37.2 could be coupled to produce a non-ω_1-saturated model of T of cardinality κ, an impossible production by 34.5. T is categorical in every uncountable power by 16.3. $\qquad\qquad\square$

Corollary 37.5 (V. Harnik, J. Ressayre). *Suppose* T *is categorical in some uncountable power,* $\mathcal{A} \subset \mathcal{B} \in \mathcal{K}(T)$ *and* $S\mathcal{A}$ *is infinite. If* $\mathcal{B}$ *is a prime model extension of* $\mathcal{A}$, *then* $\mathcal{B}$ *is a minimal model extension* $\mathcal{A}$.

Proof. Let $p \in S\mathcal{A}$ be a limit point. p is not realized in $\mathcal{B}$, since 32.6 requires $\mathcal{B}$ to be atomic over $\mathcal{A}$. Suppose $\mathcal{B}$ is not minimal over $\mathcal{A}$. Then 36.5 supplies an infinite set of indiscernibles over $\mathcal{A}$ in $\mathcal{B}$. But then 37.2 yields an uncountable unsaturated model of T, an impossible outcome by 37.4. $\quad\square$

Exercise 37.6. Find a proof of 37.5 that avoids Sec. 36.

Section 38

Minimal Generators and ω_1-Categoricity

Minimal generators are the source of a dimension theory for models of ω_1-categorical theories. (The concept of minimal generator is a reformulation of W. Marsh's concept of minimal formula [Mal].) It will be seen in the next section that every model $\mathcal{A}$ of every ω_1-categorical theory has a well-defined dimension in the same sense that every algebraically closed field has a dimension equal to the cardinality of any of its transcendence bases. Vaught's two-cardinal theorem is the key that unlocks the dimensionality of $\mathcal{A}$.

By 37.3 it is innocuous to assume T is substructure complete. For the sake of assigning Morley rank and degree to members of S_1T, add an object $\emptyset$ (called the null structure) to the category $\mathcal{K}(T)$. For each $\mathcal{A} \in \mathcal{K}(T)$ add an inclusion map $\emptyset \subset \mathcal{A}$. Let $S\emptyset$ be S_1T and define the projection $S\mathcal{A} \to S\emptyset$ by restriction: the image of $q \in S\mathcal{A}$ in $S\emptyset$ is the result of deleting all formulas of q not in the language of T. A minimal generator for T is a $p \in S_1T(= S\emptyset)$ such that rank $p = 1$ and deg $p = 1$. ACF_0 has a unique minimal generator.

Let $\{\underline{c}_i \mid i < \omega\}$ be a sequence of individual constants, none of which occur in the language of T. T^* is a finitely generated extension of T if T^* is

$$T \cup \{\underline{c}_1, \ldots, \underline{c}_n \mid F(x_1, \ldots, x_n) \in q\}$$

for some $q \in S_n T$. In symbols: $T^* = T \cup q$.

Lemma 38.1 (W. Marsh). *If T is totally transcendental, then some finitely generated extension of T has a minimal generator.*

Proof. By 31.2 T has an infinite universal domain $\mathcal{U}$. $S\mathcal{U}$ has at least one limit point, because it is infinite. So $S\mathcal{U}$ has a point of positive rank, since T is totally transcendental. It follows from 31.3 and 31.8 that there exists a $p \in S\mathcal{U}$ such that rank $p = 1$ and deg $p = 1$. Let $\mathcal{C} = \emptyset(c_1, \ldots, c_n)$ be a finitely generated substructure of $\mathcal{U}$ such that the projection of p in $S\mathcal{C}$ has rank 1 and degree 1. Let $q \in S_n T$ be the n-type realized by $\langle c_1, \ldots, c_n \rangle$. Then $T \cup q$ has a minimal generator, namely the projection of p in $S\mathcal{C}(= S_1(T \cup q))$. $\square$

Suppose $\mathcal{A} \models T$, $a \in A$ and $X \subset A$; a is algebraic over X if a realizes a 1-type of SX of rank 0. Observe that a is algebraic over X iff there is a formula $F(x)$ (in the language of $T \cup DX$) such that $\mathcal{A} \models F(\underline{a})$ and only finitely many members of $\mathcal{A}$ satisfy $F(x)$. X is algebraically independent if no $x \in X$ is algebraic over $X - \{x\}$.

Proposition 38.2. *Suppose $\mathcal{A} \models T$, p is a minimal generator for T, and $X \subset A$ is an algebraically independent set of realizations of p. Then:*

(i) *X is a set of indiscernibles in $\mathcal{A}$.*
(ii) *If a realizes p in $\mathcal{A}$, and a is not algebraic over X, then $\{a\} \cup X$ is algebraically independent.*

Proof. Let $\{x_\delta \mid \delta < \alpha\}$ be an enumeration of X. The rank and degree rules imply $\{x_\delta \mid \delta < \alpha\}$ is a Morley sequence of degree 1. (i) follows from 35.6. Let $x_\alpha = a$. Then $\{x_\delta \mid \delta \leq \alpha\}$ is a Morley sequence, hence a set of indiscernibles in $\mathcal{A}$. It follows no member of $\{x_\delta \mid \delta \leq \alpha\}$ can be algebraic over the remaining members, since otherwise a would be algebraic over X. □

Suppose T has a minimal generator p and $\mathcal{A} \models T$. X is a p-base for $\mathcal{A}$ if X is a maximal, algebraically independent set of realizations of p in $\mathcal{A}$.

Lemma 38.3 (W. Marsh). *Suppose T has a minimal generator p and $\mathcal{A} \models T$. If X and Y are p-bases for $\mathcal{A}$, then card X = card Y.*

Proof. Assume card $X \leq$ card Y. The lemma is proved for all T, all p and all $\mathcal{A}$ by induction on card X. First suppose card $X \geq \omega$. Every member of Y is algebraic over X by 38.2 (ii), so card $Y \leq$ max $(\omega,$ card $X) =$ card X, since the language of T is countable.

Now suppose card $X < \omega$. Choose $y \in Y$. By 38.2 (ii) y is algebraic over X. Since y is not algebraic over $Y - \{y\}$, there must be an $x \in X$ such that x is not algebraic over $Y - \{y\}$. Let $q \in S\{x\}$ be the unique pre-image of p of rank 1 and degree 1. Then q is a minimal generator for $T \cup p$, and $\langle \mathcal{A}, x \rangle \models T \cup p$. $X - \{x\}$ is a q-base for $\langle \mathcal{A}, x \rangle$, and $Y - \{y\}$ can be extended to some q-base y^* for $\langle \mathcal{A}, x \rangle$. But then by induction, card $(X - \{x\}) =$ card $Y^* \geq$ card $(Y - \{y\})$, and so card $X =$ card Y. □

Suppose T has a minimal generator p and $\mathcal{A} \models T$. The p-dimension of $\mathcal{A}$ is card X, where X is any p-base for $\mathcal{A}$. In symbols: p-dim $\mathcal{A} =$ card X.

Theorem 38.4 (W. Marsh). *Suppose T is ω_1-categorical and has a minimal generator p. Let $\mathcal{A}$ and $\mathcal{B}$ be models of T. Then:*

(i) *If X is a p-base for $\mathcal{A}$, then $\mathcal{A}$ is minimal and atomic over X.*

(ii) *$\mathcal{A} \approx \mathcal{B}$ iff p-dim $\mathcal{A} = p$-dim $\mathcal{B}$.*

Proof. (i) Let $F(x) \in p$ be a formula such that $F(x) \in q$ implies $q = p$ or q has rank 0. $F(x)$ exists because rank $p = 1$. Suppose $X \subset C$ and $C \prec A$. Let $F^C = \{c \mid c \in C \ \& \ C \models F(\underline{c})\}$. F^C is infinite, since $C \models T$ and p has positive rank. Clearly $F^C \subset F^A$. Suppose $a \in F^A$. If a realizes a $q \in S\emptyset$ of rank 0, then $a \in F^C$, since $C \models T$ and all models of T are algebraically closed. If a realizes p, then a is algebraic over X. Thus $F^C = F^A$, and so by 22.5, $C = A$. A is atomic over X by 32.4 and 32.6.

(ii) If $A \approx B$, then p-dim $A = p$-dim B by 38.3. Suppose p-dim $A = p$-dim B. Let X be a p-base for A and Y a p-base for B. Let $f \colon X \to Y$ be a one-one, onto map. Then

$$\langle A, x \rangle_{x \in X} \equiv \langle B, fx \rangle_{x \in X}.$$

A is prime over x by 38.4 (i) and 32.3, hence f can be extended to $g \colon A \overset{\equiv}{\to} B$. Thus $Y \subset g[A]$ and $g[A] \prec B$, and so $g[A] = B$ by 38.4 (i). □

Corollary 38.5 (W. Marsh). *If T is ω_1-categorical, A is a saturated model of T and $A \prec B$, then B is saturated.*

Proof. Let $T \cup q$ be a finitely generated extension of T such that $T \cup q$ has a minimal generator p. $T \cup q$ exists according to 38.1. Let $\langle a_1, \ldots, a_n \rangle$ realize q in A. Then $\langle A, a_1, \ldots, a_n \rangle$ and $\langle B, a_1, \ldots, a_n \rangle$ are models of $T \cup q$. $\langle A, a_1, \ldots, a_n \rangle$ has infinite p-dimension, because A is saturated. But then $\langle B, a_1, \ldots, a_n \rangle$ has infinite p-dimension. If card $B = \omega$, then B is saturated by 38.4 (ii). If card $B > \omega$, then B is saturated by 37.4. □

A is a proper elementary extension of B if $A \succ B$ and $A \neq B$. A is a prime, proper elementary extension of B if the following diagram can be completed as shown whenever C is a proper elementary extension of B.

Lemma 38.6 (M. Morley). *If $T\mathcal{B}$ is ω_1-categorical, then $\mathcal{B}$ has a prime, proper elementary extension.*

Proof. Choose $q \in S\mathcal{B}$ to have the least possible positive rank. Then there exists an $F(x) \in q$ such that if $p \neq q$ and $F(x) \in p$, then p has rank 0. Let a realize q, and let $\mathcal{A}$ be prime over $\mathcal{B}(a)$. Suppose $\mathcal{C}$ is a proper elementary extension of $\mathcal{B}$. It suffices to show q is realized in $\mathcal{C}$. Let $F^{\mathcal{B}} = \{b \,|\, b \in B \,\&\, \mathcal{B} \models F(\underline{b})\}$. Clearly $F^{\mathcal{B}} \subset F^{\mathcal{C}}$. $F^{\mathcal{B}}$ is infinite, because $\mathcal{B} \models T$ and q has positive rank. It follows from 22.5 that $F^{\mathcal{B}} \neq F^{\mathcal{C}}$. Let $c \in F^{\mathcal{C}} - F^{\mathcal{B}}$. Then c realizes q, since otherwise c realizes some $p \in S\mathcal{B}$ of rank 0 and so belongs to $\mathcal{B}$. $\qquad\square$

Morley also showed that the $\mathcal{A}$ constructed in the proof of 38.6 is minimal over $\mathcal{B}$; i.e. there is no $\mathcal{D}$ such that $\mathcal{B} \prec \mathcal{D} \prec \mathcal{A}$, $\mathcal{B} \neq \mathcal{D}$ and $\mathcal{D} \neq \mathcal{A}$. The proof of minimality will be given in Sec. 39.

Corollary 38.7 (M. Morley). *T is ω_1-categorical iff every countable model of T has a prime, proper elementary extension.*

Proof. Assume the countable models of T have the property in question. A preferred model $\mathcal{A}$ of T of cardinality ω_1 is constructed as the limit of an elementary chain. Suppose T has a prime model $\mathcal{A}_0$. $\mathcal{A}_{\delta+1}$ is a prime, proper elementary extension of $\mathcal{A}_\delta$. $\mathcal{A}_\lambda = \cup\{\mathcal{A}_\delta \,|\, \delta < \lambda\}$. Let

$$\mathcal{A} = \cup\{\mathcal{A}_\delta \,|\, \delta < \omega_1\}.$$

Suppose $\mathcal{B} \models T$ and card $\mathcal{B} = \omega_1$. With the object of showing $\mathcal{B} \approx \mathcal{A}$, decompose $\mathcal{B}$ into an elementary chain $\{\mathcal{B}_\delta \,|\, \delta < \omega_1\}$ of countable models. A function $g: \omega_1 \to \omega_1$ and a chain

$$\{f_\delta : \mathcal{A}_{g\delta} \to \mathcal{B}_\delta \,|\, 0 < \delta < \omega_1\}$$

of isomorphisms is defined by induction on δ.

1. $g0 = 0$ and $f_0 \colon \mathcal{A}_0 \to \mathcal{B}_0$.
2. $g\lambda = \cup\{g\delta \mid \delta < \lambda\}$ and $f_\lambda = \cup\{f_\delta \mid \delta < \lambda\}$.
3. $f_{\delta+1}$ is the union of a chain $\{f_\delta^\gamma \mid \gamma \leq \rho\}$ defined by induction on γ.
4. $f_{\delta+1}^0 = f_\delta$. $f_{\delta+1}^\lambda = \cup\{f_{\delta+1}^\gamma \mid \gamma < \lambda\}$.
5. Assume $f_{\delta+1}^\gamma \colon \mathcal{A}_{g\delta+\gamma} \overset{\equiv}{\to} \mathcal{B}_{\delta+1}$. If $f_{\delta+1}^\gamma$ is onto, then $\rho = \gamma$. If $f_{\delta+1}^\gamma$ is not onto, then the primeness of $\mathcal{A}_{g\delta+\gamma+1}$ provides an extension of $f_{\delta+1}^\gamma$ to $f_{\delta+1}^{\gamma+1} \colon \mathcal{A}_{g\delta+\gamma+1} \overset{\equiv}{\to} \mathcal{B}_{\delta+1}$.
6. $g(\delta + 1) = g\delta + \rho$.

Clearly $f = \cup\{f_\delta \mid \delta < \omega_1\}$ maps $\mathcal{A}$ isomorphically onto $\mathcal{B}$. It remains to show T has a prime model. Let $\mathcal{A}$ be *any* countable model of T. The above construction establishes $T \cup D\mathcal{A}$ is ω_1-categorical; by 34.4 $T \cup D\mathcal{A}$ is ω-stable. But then T is ω-stable and so has a prime model by 31.6 and 32.4. $\qquad\qquad\square$

It follows from 38.7 that ω_1-categoricity is a Π_2^1 (hence absolute) property of theories. The results of Sec. 37 imply ω_1-categoricity is a Π_1^1 notion.

Theorem 38.8 (W. Marsh, M. Morley). *Let T be ω_1-categorical, but not ω-categorical, and have a minimal generator p. Then there exists an elementary chain $\{\mathcal{A}_i \mid i \leq \omega\}$ of countable models of T such that:*

(i) $\mathcal{A}_i \not\approx \mathcal{A}_j$ *when $i < j \leq \omega$.*
(ii) $\mathcal{B} \approx \mathcal{A}_j$ *for some i whenever $\mathcal{B}$ is a countable model of T.*
(iii) *There exists an integer k such that p-dim $\mathcal{A}_i = k + i$ for all $i \leq \omega$.*
(iv) $\mathcal{A}_i$ *is homogeneous for all $i \leq \omega$.*
(v) $\mathcal{A}_\omega$ *is saturated, and $\mathcal{A}_\omega = \cup\{\mathcal{A}_i \mid i < \omega\}$.*
(vi) $\mathcal{A}_{i+1}$ *is a minimal, prime, proper elementary extension of $\mathcal{A}_i$.*

Proof. Let $\mathcal{A}_0$ be the prime model of T. Since T is not ω-categorical, $\mathcal{A}_0$ is not saturated by 18.2. Consequently p-dim $\mathcal{A}_0 < \omega$ by 38.4 (ii). Let $k = p$-dim $\mathcal{A}_0$. Fix $i < \omega$ and assume $\mathcal{A}_i$ has been defined so that p-dim $\mathcal{A}_i = k + i$. Let $q \in S\mathcal{A}_i$ be the unique pre-image of p of rank 1 and degree 1. Let b realize q and $\mathcal{A}_{i+1}$ be prime over $\mathcal{A}_i(b)$. Clearly p-dim $\mathcal{A}_{i+1}$ is at least $k + i + 1$; suppose it is greater. Then there is an $X \subset A_i$ and a $c \in A_{i+1} - A_i(b)$ such that $X \cup \{b, c\}$ is an algebraically independent set of realizations of p. In addition the 1-type $r \in S\mathcal{A}_i(b)$ realized by c is isolated by 32.6. Let $V \subset S\mathcal{A}_i$ be a clopen set whose only members are q and points of rank 0. V must be infinite, since otherwise q would be isolated, realized in $\mathcal{A}_i$, and of rank 0. Let V_r be the pre-image of V in $S\mathcal{A}_i(b)$. Consider $V_r - \{r\}$: closed since r is isolated; infinite since V is infinite; contains only rank 0 points. Thus $V_r - \{r\}$ is a closed, infinite set without a limit point, and so dim $\mathcal{A}_{i+1} = k + i + 1$.

Let $\mathcal{A}_\omega = \cup \{\mathcal{A}_i \,|\, i < \omega\}$. Then p-dim $\mathcal{A}_\omega = \omega$; hence $\mathcal{A}_\omega$ is saturated by 38.4 (ii).

Let $\mathcal{A}$ be a model of T of finite p-dimension in the hope of showing $\mathcal{A}$ is homogeneous. Assume $f \colon X \to Y$ is a one-one onto map such that

$$\langle \mathcal{A}, x \rangle_{x \in X} \equiv \langle \mathcal{A}, fx \rangle_{x \in X}.$$

Define $T_1 = T \cup DX$. Let $q \in SX$ be the unique pre-image of p of rank 1 and degree 1. Then q is a minimal generator for T_1. Let Y be a q-base for $\langle \mathcal{A}, x \rangle_{x \in X}$ and Z be a q-base for $\langle \mathcal{A}, fx \rangle_{x \in X}$. Assume card $Y \leq$ card Z. Let $g \colon Y \to Z$ be one-one. $\langle \mathcal{A}, x \rangle_{x \in X}$ is prime over Y be 38.4 (i) and 32.3, hence g can be extended to $h \colon \langle \mathcal{A}, x \rangle_{x \in X} \overset{\equiv}{\to} \langle \mathcal{A}, fx \rangle_{x \in X}$. Thus f can be extended to $h \colon \mathcal{A} \overset{\equiv}{\to} \mathcal{A}$. Let $W \subset h[A]$ be a p-base for $h[\mathcal{A}]$. Then W is also a p-base for $\mathcal{A}$ by 38.3, since $h[\mathcal{A}]$ and $\mathcal{A}$ have the same p-dimension and any p-base for $h[\mathcal{A}]$ can be extended to a p-base for $\mathcal{A}$. But then $h[A] = A$ by 38.4 (i). (Note that the above argument did not assume card $X <$ card A.)

Suppose $\mathcal{A} \prec \mathcal{C} \prec \mathcal{B}$ and $\omega > p\text{-dim } \mathcal{B} = 1 + p\text{-dim } \mathcal{A}$. Then either $p\text{-dim } \mathcal{C} = p\text{-dim } \mathcal{B}$ or $p\text{-dim } \mathcal{C} = p\text{-dim } \mathcal{A}$. It follows from 38.4 (i) that $\mathcal{C} = \mathcal{B}$ or $\mathcal{C} = \mathcal{A}$. $\qquad\square$

In the next section all the conclusions of 38.8 save (iii) will be proved without the assumption that T has a minimal generator.

Exercise 38.9. Suppose T_1 is an ω_1-categorical theory with a minimal generator, $\mathcal{A}$ is an unsaturated model of T, and f: $\mathcal{A} \overset{\equiv}{\to} \mathcal{A}$. Show f is onto. (cf. Exercise 39.14.)

Exercise 38.10. Suppose $T\mathcal{A}$ is ω_1-categorical. Show $\mathcal{A}$ is saturated iff $\mathcal{A}$ contains an infinite set of indiscernibles.

Section 39

The Baldwin–Lachlan Theorem

Let T^* be a finitely generated extension of T as defined in Sec. 38. T^* is principal if $T^* = T \cup q$, where q is a principal n-type of T. The next lemma is the essential modification of 38.1 needed to develop a dimension theory for ω_1-categorical theories that lack minimal generators.

Lemma 39.1 (J. Baldwin, A. Lachlan). *If T is ω_1-categorical, then some principal, finitely generated extension of T has a minimal generator.*

Proof. Let $\mathcal{A}$ be the prime model of T, and let $p \in S\mathcal{A}$ have the least possible positive rank and the least possible degree in that rank. If rank $p = 1$ and degree $p = 1$, then the projection of p in $S\mathcal{E}$ has rank 1 and degree 1 for some finitely generated $\mathcal{E} \subset \mathcal{A}$. But then $\mathcal{E} = \emptyset(e_1, \ldots, e_n)$, $\langle e_1, \ldots, e_n \rangle$ realizes some principal n-type $q \in S_n t$, and $T \cup q$ has a minimal generator.

Suppose rank $p > 1$ or deg $p > 1$ with the intent of showing T is not ω_1-categorical. Then there exists a $\mathcal{B} \succ \mathcal{A}$ such that p has distinct pre-images, p_1 and p_2, of positive rank in $S\mathcal{B}$. Let $F(x)$ be a formula (in the language of $T \cup D\mathcal{A}$) such that $F(x) \in p$, and such that $F(x) \in q$ implies $q = p$ or rank $q = 0$. $F(x)$ exists because no $q \in S\mathcal{A}$ has lower positive rank than p. Let $G(x)$ be a formula (in the language of $T \cup D\mathcal{B}$) such

172

that $G(x) \in p_1$ and $\sim G(x) \in p_2$. By 32.11 $G(x)$ & $F(x)$ has infinitely many realizations in $\mathcal{B}$, since otherwise rank $p_1 = 0$. On similar grounds $\sim G(x)$ & $F(x)$ has infinitely many realizations in $\mathcal{B}$. For the sake of notational simplicity, suppose b is the only member of $B{-}A$ mentioned in $G(x)$; i.e. $G(x)$ is $G(x, \underline{b})$. For each $a \in A$, 32.11 implies one of the formulas

$$G(x, \underline{a}) \ \& \ F(x), \quad \sim G(x, \underline{a}) \ \& \ F(x),$$

must have only finitely many realizations in $\mathcal{A}$, since there is only one q of positive rank, namely p, such that $F(x) \in q$. Suppose there were an integer n such that for every $a \in A$, either

$$G(x, \underline{a}) \ \& \ F(x) \quad \text{or} \quad \sim G(x, \underline{a}) \ \& \ F(x)$$

had at most n realizations in $\mathcal{A}$. Then either $G(x)$ & $F(x)$ or $\sim G(x)$ & $F(x)$ would have at most n realizations in $\mathcal{B}$, since $\mathcal{B} \succ \mathcal{A}$. It follows there is a formula $H(x, y)$ (either $G(x, y)$ & $F(x)$ or $\sim G(x, y)$ & $F(x)$) such that for each n there is an $a \in A$ such that $H(x, \underline{a})$ has at least n, but not infinitely many, realizations in $\mathcal{A}$. Note that the existence of $H(x, y)$ is a consequence of the ability of $S\mathcal{A}$ to "approximate" the splitting of p in $S\mathcal{B}$, an ability derived from the fact that $\mathcal{A} \prec \mathcal{B}$.

As in the proof of 22.1 it is possible to write down axioms Q that describe a pair $\langle \mathcal{D}, \mathcal{C} \rangle$ of structures such that $\mathcal{D} \supset \mathcal{C}$. Q says:

(i) $\mathcal{C} \models T,\ \mathcal{C} \prec \mathcal{D},\ \mathcal{C} \neq \mathcal{D}$.

(ii) $K^{\mathcal{C}} = K^{\mathcal{D}}$, where $K(x)$ is $H(x, \underline{c})$ and $\underline{c}$ is an individual constant not occurring in (i).

(iii) $\underline{c} \in \mathcal{C}$.

(iv) For each n, $H(x, \underline{c})$ has at least n realizations in $\mathcal{C}$.

Any finite subset of Q can be satisfied in $\langle \mathcal{B}, \mathcal{A} \rangle$: Choose $c \in A$ so that $H(x, \underline{c})$ has a sufficiently large *finite* number of realizations in $\mathcal{A}$. Then $K^{\mathcal{B}} = K^{\mathcal{A}}$, since $\mathcal{B} \succ \mathcal{A}$ and $K^{\mathcal{A}}$ is finite. By 22.5 T is not ω_1-categorical. $\qquad\square$

T has a nonprincipal minimal generator if T has a minimal generator p such that p is not an isolated point of $S_1 T$.

Lemma 39.2. *If T is ω_1-categorical, but not ω-categorical, then some principal, finitely generated extension of T has a nonprincipal minimal generator.*

Proof. By 39.1 there is an n, a principal $q \in S_n T$, and a $p \in S_1(T \cup q)$ such that rank $p = 1$ and deg $p = 1$. $T \cup q$ is not ω-categorical because q is realized in every model of T. Let $\langle \mathcal{A}, a_1, \ldots, a_n \rangle$ be the prime model of $T \cup q$. By 38.8 $\langle \mathcal{A}, a_1, \ldots, a_n \rangle$ has finite p-dimension. Let $\{x_1, \ldots, x_k\}$ be a p-base for $\langle \mathcal{A}, a_1, \ldots, a_n \rangle$.

Let $r \in S_{n+k} T$ be the isolated point realized by $\langle a_1, \ldots, a_n, x_1, \ldots, x_k \rangle$ in $\mathcal{A}$. Let p^* be the unique pre-image of p in $S\emptyset(a_1, \ldots, a_n, x_1, \ldots, x_k)$ of rank 1 and degree 1. Then p^* is a minimal generator for $T \cup r$. p^* is not an isolated point of $S_1(T \cup r)$; if it were, it would be realized in $\mathcal{A}$, and then $\{x_1, \ldots, x_k\}$ would not be a p-base for $\langle \mathcal{A}, a_1, \ldots, a_n \rangle$.   □

Let T be an ω_1-categorical, but not ω-categorical, theory. Lemma 39.2 suggests the following notion of dimension for models of T. Choose an n, a principal $q \in S_n T$, and a $p \in S_1(T \cup q)$ such that p is a nonprincipal minimal generator for $T \cup q$. Let $\mathcal{A}$ be any model of T. Since q is principal, there exist $a_1, \ldots, a_n \in A$ such that $\langle \mathcal{A}, a_1, \ldots, a_n \rangle$ is a model of $T \cup q$. Define the

$$p(\bmod \langle a_1, \ldots, a_n \rangle)\text{-dimension of } \mathcal{A}$$

to be the p-dimension of $\langle \mathcal{A}, a_1, \ldots, a_n \rangle$. If the $p(\bmod \langle \mathcal{A}, a_1, \ldots, a_n \rangle)$-dimension remains constant as $\langle a_1, \ldots, a_n \rangle$ ranges over all realizations of q in $\mathcal{A}$, then the $p(\bmod q)$-dimension of $\mathcal{A}$ is said to be well defined. If $\mathcal{A}$ is saturated, then the $p(\bmod q)$-dimension of $\mathcal{A}$ is well defined and equal to card $\mathcal{A}$. With the aid of Vaught's two-cardinal theorem it will be shown that the $p(\mathrm{mod}\ q)$-dimension of $\mathcal{A}$ is well defined for every model $\mathcal{A}$ of T. Then it will be possible to repeat the proof of 38.8 without assuming T has a minimal generator.

Proposition 39.3. *Suppose T has a nonprincipal minimal generator p, and $C \in \mathcal{K}(T)$. Let $q \in SC$ be the unique pre-image of p of rank 1 and degree 1. Then q is a nonprincipal minimal generator for $T \cup DC$.*

Proof. Let $U \subset S_1 T$ be a clopen neighborhood of p that contains only points of rank 0 save for p. U is infinite since p is nonprincipal. Let $V \subset SC$ be the pre-image of U. By the rank and degree rules, q is the sole member of V of positive rank. If q were isolated, then $V - \{q\}$ would be an infinite closed set without a limit point. □

The next three propositions are needed for the proof of Theorem 39.8.

Proposition 39.4. *Let T be a theory with a nonprincipal minimal generator p. If b realizes p or is algebraic (i.e. realizes a $q \in S_1 T$ of rank 0), and c realizes a principal 1-type of T, then c realizes an atom over b.*

Proof. First allow b to be algebraic. Then b realizes an atom over c; $\langle b, c \rangle$ realizes a principal 2-type of T by 32.5; and so c realizes an atom over b.

Now suppose b realizes p. Then b cannot realize an atom over c; if b did, then $\langle b, c \rangle$ would realize a principal 1-type of T and p would be principal. It follows from the rank and degree rules that b realizes a 1-type q over c of rank 1 and degree 1. Let $H(\underline{c}, x)$ be a formula such that $H(\underline{c}, x) \in q$, and such that $H(\underline{c}, x) \in r$ implies $r = q$ or rank $r = 0$. Let $F(x)$ generate the principal 1-type of T satisfied by c. The 2-type of $S_2 T$ realized by $\langle c, b \rangle$

is completely described by: $F(c)$, $H(c,b)$ and b is not algebraic over c. It follows that the 1-type s realized by c over b is given by: $F(x)$, $H(x,\underline{b})$ and $\underline{b}$ is not algebraic over x. To see that s is generated by $F(x)$ & $H(x,\underline{b})$ it is enough to see there cannot be a c^* such that $F(c^*)$ holds and b is algebraic over c^*. If there were such a c^*, then $\langle b, c^* \rangle$ would realize principal 1-type of T, and p would be principal. $\square$

Proposition 39.5. *Let T be a theory with a nonprincipal minimal generator p. If every member of $\{b_1, \ldots, b_m\}$ realizes p or is algebraic, and $\langle c_1, \ldots, c_n \rangle$ realizes a principal n-type of T, then $\langle c_1, \ldots, c_n \rangle$ realizes an atom over $\{b_1, \ldots, b_m\}$.*

Proof. By induction on n. The case of $n = 1$ is handled by induction on m. Suppose c realizes a principal 1-type of T. Then by induction c realizes an atom over $\{b_1, \ldots, b_{m-1}\}$. Let $p_m \in S\emptyset(b_1, \ldots, b_{m-1})$ be the unique pre-image of p of rank 1 and degree 1. By 39.3 p_m is a minimal generator for $T \cup D\emptyset(b_1, \ldots, b_{m-1})$. Either b_m is algebraic over $\{b_1, \ldots, b_{m-1}\}$ or realizes p_m. So by 39.4 c realizes an atom over $b_1, \ldots, b_m$.

Assume $n > 1$. Let $p_n \in S\emptyset(c_n)$ be the unique pre-image of p of rank 1 and degree 1. By 39.3 p_n is a minimal generator for $T \cup D\emptyset(c_n)$. Every member of $\{b_1, \ldots, b_m\}$ is algebraic over c_n or realizes p_n. By 32.7 $\langle c_1, \ldots, c_{n-1} \rangle$ realizes a principal $(n-1)$-type of $T \cup D\emptyset(c_n)$. By induction on n: $\langle c_1, \ldots, c_{n-1} \rangle$ realizes an atom over $\{c_n, b_1, \ldots, b_m\}$ and c_n realizes an atom over $\{b_1, \ldots, b_m\}$. So by 32.5, $\langle c_1, \ldots, c_n \rangle$ realizes an atom over $\{b_1, \ldots, b_m\}$. $\square$

Proposition 39.6. *Let T be a theory with a nonprincipal minimal generator p. Suppose every member of B realizes p or is algebraic. If $B \subset C = \{c_0, \ldots, c_m\}$, and every $c_i \in C - B$ realizes an atom over $\{c_j \mid j < i\}$, then $C - B$ realizes an atom over B.*

Proof. By induction on card $(C - B)$. Let $c_k \in C - B$ have the property that $c_j \in B$ whenever $k < j \leq m$. c_k realizes an atom over $\{c_0, \ldots, c_{k-1}\}$. Let $p_k \in S\emptyset(c_0, \ldots, c_{k-1})$ be the unique pre-image of p of rank 1 and degree 1. By 39.3 p_k is a nonprincipal minimal generator for $T \cup D\emptyset(c_0, \ldots, c_{k-1})$. Every member of $\{c_j \mid k < j \leq m\}$ realizes p_k or is algebraic over $\{c_0, \ldots, c_{k-1}\}$. It follows from 39.5 that c_k realizes an atom over $C - \{c_k\}$. By induction $(C - B) - \{c_k\}$ realizes an atom over B. But then 32.5 implies $C - B$ realizes an atom over B. $\qquad\square$

Let $F(x)$ be a formula in the language of T. $\langle \mathcal{B}, \mathcal{A} \rangle$ is a Vaughtian pair for $(T, F(x))$ if $\mathcal{B} \succ \mathcal{A}$, $\mathcal{A} \models T$, $\mathcal{A} \neq \mathcal{B}$, $F^{\mathcal{A}}$ is infinite and $F^{\mathcal{A}} = F^{\mathcal{B}}$. According to 22.5 $(T, F(x))$ cannot have a Vaughtian pair when T is ω_1-categorical. It is not difficult to devise a set K of sentences such that $T \cup K$ is consistent iff $(T, F(x))$ has a Vaughtian pair, but some finesse is required to design a K that makes clear on proof theoretic grounds — reminiscent of Herbrand's theorem — just how it happens that $(T, F(x))$ has a Vaughtian pair. K is needed for the proof of the compactness phenomenon expressed by 39.8.

Let $\underline{c}, \underline{c}_0, \underline{c}_1, \underline{c}_2, \ldots$ be a sequence of distinct individual constants, none of which occur in T. The design of K is inspired by the Henkin style proof of 7.1. Let $\{G_i(x)\}$ be an enumeration of all formulas (in the language of T with $\underline{c}, \underline{c}_0, \underline{c}_1, \underline{c}_2, \ldots$ adjoined) whose sole free variable, if any, is x. Let $h : w \to w$ be a strictly increasing function such that $k < hi$ whenever $j \leq i$ and $\underline{c}_k$ occurs in $G_j(x)$. The Henkin constants are $\{\underline{c}_{hi} \mid i < \omega\}$. The Henkin axiom attached to $\underline{c}_{hi}$ is

$$(Ex)G_i(x) \to G_i(\underline{c}_{hi}).$$

Let A, B and C be disjoint sets of Henkin constants that satisfy closure conditions (a), (b) and (c). $A \cup B \cup C$ is intended to name the elements of a model $\mathcal{B}_1$ of T; $A \cup B$ corresponds to the universe of some $\mathcal{B}_0 \prec \mathcal{B}_1$; and $\mathcal{A}$ to $F^{\mathcal{B}_0}$.

(a) $\underline{c}_{hi} \in A$ if $G_i(x)$ is of the form

$$F(x) \ \& \ [(Ex)(F(x) \ \& \ H(x)) \to H(x)],$$

where the individual constants in $H(x)$ belong to $T \cup \{\underline{c}\} \cup A \cup B \cup C$.

(b) $\underline{c}_{hi} \in B$ if $G_i(x)$ is of the form

$$\sim F(x) \ \& \ [(Ex)(\sim F(x) \ \& \ H(x)) \to H(x)],$$

where the individual constants occurring in $H(x)$ belong to $T \cup A \cup B$.

(c) $\underline{c}_{hi} \in C$ if $G_i(x)$ has the same form as in (b), the individual constants occurring in $H(x)$ belong to $T \cup \{\underline{c}\} \cup A \cup B \cup C$, and some individual constant from $\{\underline{c}\} \cup C$ occurs in $H(x)$.

Let Δ be the set of all Henkin axioms attached to members of $A \cup B \cup C$. Then K is

$$\Delta \cup \{\sim F(\underline{c})\} \cup \{\underline{c} \neq \underline{c}_{hi} \mid \underline{c}_{hi} \in B\}.$$

Lemma 39.7. *Suppose T is a theory such that $F^{\mathcal{A}}$ is infinite whenever $\mathcal{A} \models T$. Then $(T, F(x))$ has a Vaughtian pair iff $T \cup K$ is consistent.*

Proof. First suppose $\langle \mathcal{B}_1, \mathcal{B}_0 \rangle$ is a Vaughtian pair for $(T, F(x))$. Evaluate $\{\underline{c}\} \cup \{\underline{c}_{hi} \mid i < \omega\}$ in $\mathcal{B}_1$ by induction on i. Let the value of $\underline{c}$ be any $c \in B_1 - B_0$. Fix i and assume $\underline{c}_{hj}$ has been evaluated for all $j < i$. Thus every individual constant occurring in $G_i(x)$ has been evaluated. Choose c_{hi} to satisfy the Henkin axiom attached to $\underline{c}_{hi}$; in addition, choose $c_{hi} \in B_0$ if $\underline{c}_{hi} \in A \cup B$. The desired value of $\underline{c}_{hi}$ exists when $\underline{c}_{hi} \in A \cup B$, because $F^{\mathcal{B}_1} = F^{\mathcal{B}_0}$ and $\mathcal{B}_0 \prec \mathcal{B}_1$. Then $\langle \mathcal{B}_1, c, c_{hi} \rangle_{i < \omega}$ is a model of $T \cup K$.

Now suppose $T \cup K$ is consistent. Let $\langle \mathcal{D}, c, c_{hi} \rangle_{i < \omega}$ be a model of $T \cup K$. Define $B_0 = \{c_{hi} \mid \underline{c}_{hi} \in A \cup B\}$ and $B_1 = \{c_{hi} \mid \underline{c}_{hi} \in A \cup B \cup C\}$. Define the relations and functions of $\mathcal{B}_i (i = 0, 1)$

by restricting the relations and functions of $\mathcal{D}$ to $\mathcal{B}_i$. If all the individual constants occurring in $G_i(x)$ name members of B_1, and the Henkin axiom for $G_i(x)$ is true in $\mathcal{D}$, then it is true in $\mathcal{B}_1$. Consequently $\mathcal{B}_1 \prec \mathcal{D}$ by the Skolem hull argument of 11.2. Similarly $\mathcal{B}_0 \prec \mathcal{B}_1$. Each element of $F^{\mathcal{B}_1}$ is named by some member of A, so $F^{\mathcal{B}_0} = F^{\mathcal{B}_1}$. $c \in B_1$ since c is named by some member of C. c is not named by any member of $A \cup B$, because $\mathcal{D}$ satisfies $\sim F(\underline{c})$ and $\underline{c} \neq \underline{c}_{hi}$ for all $\underline{c}_{hi} \in B$. Hence $B_1 \neq B_0$. $\qquad\square$

Suppose T satisfies the hypothesis of Theorem 39.8. Let $\mathcal{D}$ be a model of T, X a p-base for $\mathcal{D}$, and $c \in D$. Then c realizes an atom over X by 38.4 (i). Of course c realizes an atom over some finite $Y \subset X$. Define the rank of c in $\mathcal{D}$ to be the least value of card Y as X ranges over all p-bases for $\mathcal{D}$. (If T is the theory of algebraically closed fields of characteristic 0, then the rank of c in $\mathcal{D}$ is at most 1, since c is either algebraic or the first member of a transcendence base for $\mathcal{D}$.) Theorem 39.8 says that the rank of c in $\mathcal{D}$ is bounded by some finite n as $\mathcal{D}$ ranges over all models of T and c ranges over all $c \in D$. The existence of n is one of the most pleasing compactness phenomena in all of model theory.

Theorem 39.8 (J. Baldwin, A. H. Lachlan). *Suppose T is ω_1-categorical and has a nonprincipal minimal generator p. Then there exists an integer n with the following property: for every model $\mathcal{D}$ of T and every $c \in D$, there is a p-base X for $\mathcal{D}$ such that c realizes an atom over some subset of X of cardinality at most n.*

Proof. Let $F(x) \in p$ be a formula such that $F(x) \in q \in S_1 T$ implies $q = p$ or rank $q = 0$. By 22.5 and 39.7, $T \cup K$ is inconsistent. Choose n so that K_n is inconsistent and consists of:

(i) T;

(ii) the Henkin axiom attached to $\underline{c}_{hi}$ if $i < n$;

(iii) $\sim F(\underline{c})$;

(iv) $\underline{c} \neq \underline{c}_{hi}$ if $i < n$ and $\underline{c}_{hi} \in B$.

The desired X is obtained by evaluating $\{\underline{c}\} \cup \{\underline{c}_{hi} \mid i < n\}$ in $\mathcal{D}$.

(1) Evaluate $\underline{c}$ as c.

(2) Fix $i < n$ and assume $\underline{c}_{hj}$ has been evaluated for all $j < i$. Thus every constant occurring in $G_i(x)$ has been evaluated. Choose c_{hi} to satisfy the Henkin axiom attached to $\underline{c}_{hi}$ in $\mathcal{D}$. Let $I_i = \{c_{hj} \mid j < i \ \& \ \underline{c}_{hj} \in A \cup B\}$. If $\underline{c}_{hi} \in B$, require c_{hi} to have the additional property of realizing an atom over I_i. Such a choice is possible by 32.1 (i) and 32.2, because every $\underline{c}_{hj}$ that occurs in $G_i(x)$ also occurs in I_i when $\underline{c}_{hi} \in B$.

Let $Y = \{c_{hi} \mid \underline{c}_{hi} \in A \ \& \ i < n\}$ and $Z = \{c_{hi} \mid \underline{c}_{hi} \in B$ and $i < n\}$. It is clear that every member of Y satisfies $F(x)$, and hence realizes p or is algebraic. It follows from 39.6, and from the requirement imposed on c_{hi} when $\underline{c}_{hi} \in B$, that Z realizes an atom over Y. Let X_0 be a maximal set of algebraically independent realizations of p in Y. Then Z realizes an atom over X_0 by 38.2 (ii) and 32.5. The inconsistency of K_n implies: $c \in Z$ or c satisfies $F(x)$ in $\mathcal{D}$. In either event c realizes an atom over an algebraically independent set of realizations of p of cardinality at most n. $\qquad\square$

Corollary 39.9. *Suppose T is ω_1-categorical and has a non-principal minimal generator p. Then there exists an n with the following property: for every $s > 0$, every model $\mathcal{D}$ of T, and every $\langle d_1, \ldots, d_s \rangle \in D^s$, there is a p-base X for $\mathcal{D}$ such that $\langle d_1, \ldots, d_s \rangle$ realizes an atom over some subset of X of cardinality at most ns.*

Proof. By induction on s. Choose K_n as in the proof of 39.8. Assume $s > 1$. Then by induction there is a set Y of algebraically independent realizations of p such that $\langle d_1, \ldots, d_{s-1} \rangle$ realizes an atom over Y and card $Y \leq n(s-1)$. Let T^* be $T \cup q$, where $q \in S_{s-1}T$ is the type realized by $\langle d_1, \ldots, d_{s-1} \rangle$ in $\mathcal{D}$. Let $p* \in S_1(T \cup q)$ be the unique pre-image of p of rank 1 and degree 1. Then by 39.3 $p*$ is a nonprincipal minimal generator for T^*. Let $F(x)$ be the formula mentioned in the first sentence of the proof of 39.8. Then $F(x) \in p^*$, and $F(x) \in r \in S_1 T^*$ implies $r = p^*$ or rank $r = 0$. Clearly $T^* \cup K_n$ is inconsistent.

Repeat the argument of 39.8 with two changes: evaluate $\underline{c}$ as d_s; if $\underline{c}_{hi} \in B$, require c_{hi} to realize an atom over $I_i \cup \{d_1, \ldots, d_{s-1}\}$. Then d_s realizes an atom over $X \cup \{d_1, \ldots, d_{s-1}\}$, where X is a set of algebraically independent realizations of $p*$ and card $X \leq n$. By 39.5: d_s realizes an atom over $Y \cup X \cup \{d_1, \ldots, d_{s-1}\}$, and $\langle d_1, \ldots, d_{s-1} \rangle$ realizes an atom over $Y \cup X$. So $\langle d_1, \ldots, d_s \rangle$ realizes an atom over $Y \cup X$. □

Corollary 39.10. *Let T be ω_1-categorical, q a principal m-type of T, and p a nonprincipal minimal generator of $T \cup q$. Then every model of T has a well-defined $p(\mathrm{mod}\ q)$-dimension.*

Proof. Let $\mathcal{A}$ be a model of T, and let $\langle a \rangle$ and $\langle b \rangle$ be realizations of q in $\mathcal{A}$ so that

$$p\text{-dim}\langle \mathcal{A}, \langle a \rangle \rangle < p\text{-dim}\langle \mathcal{A}, \langle b \rangle \rangle.$$

Both of the above dimensions are finite; if not, $\mathcal{A}$ is saturated by 38.8, and consequently has a well-defined $p(\mathrm{mod}\ q)$-dimension equal to its cardinality.

Let i be the smaller of the above two p-dimensions, and j the larger. The prime model of $T \cup q$ has p-dimension 0, since p is nonprincipal. $\langle \mathcal{A}, \langle b \rangle \rangle$ is a model of $T \cup q$ of p-dimension j. It follows from 38.8 that there is an $\mathcal{A}_0$ such that $\langle \mathcal{A}_0, \langle b \rangle \rangle \prec \langle \mathcal{A}, \langle b \rangle \rangle$ and the p-dimension of $\langle \mathcal{A}_0, \langle b \rangle \rangle$ is i. $\langle \mathcal{A}_0, \langle b \rangle \rangle \approx \langle \mathcal{A}, \langle a \rangle \rangle$, since $\langle \mathcal{A}_0, \langle b \rangle \rangle$ and $\langle \mathcal{A}, \langle a \rangle \rangle$ are models of $T \cup q$ of the same p-dimension. $\mathcal{A}_0 \neq \mathcal{A}$ because the $p(\mathrm{mod}\ \langle b \rangle)$-dimension of $\mathcal{A}_0$ is less than the $p(\mathrm{mod}\ \langle b \rangle)$-dimension of $\mathcal{A}$.

Thus $\mathcal{A}_0 \prec \mathcal{A}$, $\mathcal{A}_0 \approx \mathcal{A}$ and $\mathcal{A}_0 \neq \mathcal{A}$. A simple induction shows that for each t there is a realization $\langle c \rangle$ of q in $\mathcal{A}$ such that the $p(\bmod \langle c \rangle)$-dimension of $\mathcal{A}$ is at least t. Fix t and suppose the previous assertion holds. Since $\mathcal{A}_0 \approx \mathcal{A}$, there is a realization $\langle c_0 \rangle$ of q in $\mathcal{A}_0$ such that the $p(\bmod \langle c_0 \rangle)$-dimension of $\mathcal{A}_0$ is at least t. Since $\mathcal{A}_0 \prec \mathcal{A}$ and $\mathcal{A}_0 \neq \mathcal{A}$, the $p(\bmod \langle c_0 \rangle)$-dimension of $\mathcal{A}$ is at least $t + 1$.

Choose $\langle c \rangle$ to realize q in $\mathcal{A}$ so that the p-dimension of $\langle \mathcal{A}, \langle c \rangle \rangle$ is at least $1 + n(i + m)$, where n is the integer supplied by 39.9 for models of $T \cup q$. Let Y be a p-base for $\langle \mathcal{A}, \langle a \rangle \rangle$. By 39.9 there is a p-base X for $\langle \mathcal{A}, \langle c \rangle \rangle$ such that $Y \cup \langle a \rangle$ realizes an atom (in $\langle \mathcal{A}, \langle c \rangle \rangle$) over some subset X_0 of X of cardinality at most $n(i + m)$. $\mathcal{A}$ is atomic over $Y \cup \langle a \rangle$ by 38.4 (i), hence $\mathcal{A}$ is atomic over $X_0 \cup \langle c \rangle$. But then the p-dimension of $\langle \mathcal{A}, \langle c \rangle \rangle$ is at most $n(i + m)$ by 38.4 (i). $\qquad \square$

Corollary 39.11 (J. Baldwin, A. H. Lachlan). *If T is ω_1-categorical, but not ω-categorical, then the number of countable models of T is ω.*

Proof. By 39.2 there is an m, a principal $q \in S_m T$, and a nonprincipal $p \in S_1(T \cup q)$ such that p is a minimal generator for $T \cup q$. Since T is not ω-categorical, neither is $T \cup q$. By 38.8: the number of countable models of $T \cup q$ is ω, and any two nonisomorphic models of $T \cup q$ have different p-dimensions. It follows from 39.10 that any two nonisomorphic models of $T \cup q$ are nonisomorphic models of T. $\qquad \square$

Corollary 39.10 allows the assignment of a definite dimension to every model of every ω_1-categorical theory T. Suppose $\mathcal{A} \models T$. If $\mathcal{A}$ is saturated, then the dimension of $\mathcal{A}$ is card $\mathcal{A}$. Suppose T has a model which is not saturated. Then T is not ω-categorical. Choose an m, a principal $q \in S_m T$, and a nonprincipal $p \in S_1(T \cup q)$ such that p is a minimal generator for $T \cup q$. Let $\mathcal{A}$ be an unsaturated model of T. Then according to 39.10, $\mathcal{A}$ has a well-defined, finite $p(\bmod q)$-dimension. The value of $p(\bmod q)$-dim $\mathcal{A}$ does not depend on the choice of $\langle m, q, p \rangle$. Every choice of $\langle m, q, p \rangle$ gives rise to some ω-tower of countable models as described in the conclusion of 38.8. Each such tower begins with the prime model of T; and by induction on i, all such towers have the same model on the i-th level. The induction turns on the fact that the $(i + 1)$-th model in each tower is a minimal, proper model extension of the i-th model in that tower. So there is no risk in defining dim $\mathcal{A}$ to be the $p(\text{mod } q)$-dimension of $\mathcal{A}$ for any choice of $\langle m, q, p \rangle$.

Corollary 39.12 (J. Baldwin, A. H. Lachlan). *If T is ω_1-categorical, then every model of T is homogeneous.*

Proof. Let $\mathcal{A}$ be a model of T and $f \colon X \to Y$ an onto map such that

$$\langle \mathcal{A}, x \rangle_{x \in X} \equiv \langle \mathcal{A}, fx \rangle_{x \in X}.$$

Assume $\mathcal{A}$ is not saturated. By 39.2 there is an m, a principal $q \in S_m(T \cup DX)$, and a nonprincipal $p \in S_1(T \cup DX \cup q)$ such that p is a minimal generator for $T \cup DX \cup q$. Extend f so that $\langle a \rangle$ realizes q in $\langle \mathcal{A}, x \rangle_{x \in X}$ and $\langle fa \rangle$ realizes q in $\langle \mathcal{A}, fx \rangle_{x \in X}$. Let Y be a p-base for $\langle \mathcal{A}, x, \langle a \rangle \rangle$ and Z a p-base for $\langle \mathcal{A}, fx, \langle fa \rangle \rangle$. Assume card $Y \leq$ card Z. Choose a one-one into $g \colon Y \to Z$, $\langle \mathcal{A}, x, \langle a \rangle \rangle$ is prime over Y by 38.4 (i) and 32.3, hence g can be extended to

$$h \colon \langle \mathcal{A}, x, \langle a \rangle \rangle_{x \in X} \overset{\equiv}{\to} \langle \mathcal{A}, fx, \langle fa \rangle \rangle_{x \in X}.$$

Clearly h extends f, and $h \colon \mathcal{A} \overset{\equiv}{\to} \mathcal{A}$.

By 39.2 there is an m^*, a principal $q^* \in Sm^*T$, and a nonprincipal $p* \in S_1(T \cup q*)$ such that p^* is a minimal generator for $T \cup q*$. Let $\langle b \rangle$ realize $q*$ in $\mathcal{A}$, and let $W \subset h[A]$ be a $p*$-base for $\langle h[\mathcal{A}], \langle hb \rangle \rangle$. W is finite since $\mathcal{A}$ is not saturated. By 39.10 $h[\mathcal{A}]$ and $\mathcal{A}$ have the same finite $p*(\text{mod } q*)$-dimension, so W is also a p^*-base for $\langle \mathcal{A}, \langle hb \rangle \rangle$. But then $h[A] = A$ by 38.4 (i). $\qquad \square$

A. H. Lachlan [La1] has improved 39.11 by showing: if T is totally transcendental, but not ω-categorical, then T has infinitely many models. His improvement does not yield Corollary 39.12.

Exercise 39.13. Suppose T is ω_1-categorical but not ω-categorical. Fix $n > 0$. Show T has an unsaturated model that realizes every n-type of T.

Exercise 39.14. Suppose $T\mathcal{A}$ is ω_1-categorical, $\mathcal{A}$ is not saturated, and $f\colon \mathcal{A} \overset{\equiv}{\to} \mathcal{A}$. Show f is an isomorphism.

Exercise 39.15. Suppose $T\mathcal{D}$ is ω_1-categorical, $c \in D$, and $\mathcal{D}$ is prime over c. Let $q \in S\emptyset(= S_1 T)$ be the 1-type realized by c. Show rank $q \geq \dim \mathcal{D}$.

Section 40

Differential Fields of Characteristic 0

The language of differential fields is the language of fields augmented by a 1-place function symbol D. The theory of differential fields of characteristic 0 (DF_0) is the theory of fields of characteristic 0 increased by two axioms that relate to the derivative D:

$$D(x + y) = Dx + Dy$$
$$D(x \cdot y) = xDy + yDx.$$

DF_0 is a universal theory, hence a promising candidate for the treatment accorded in Sec. 17 to the theory of ordered fields by Blum's criterion (17.2) for detecting model completions. The notion of differential fields goes back to Ritt [Ri1], but the theory of differentially closed fields is a recent invention of A. Robinson [Ro1]. Robinson proved somewhat indirectly that DF_0 has a model completion, and then defined DCF_0 to be the unique (by 12.1) model completion of DF_0. Subsequently Blum [Bl1] found simple axioms for DCF_0, simple in that they make no mention of differential polynomials in more than one variable.

Let $\mathcal{A}$ be a differential field of characteristic 0. A differential polynomial over $\mathcal{A}$ in the variables x_i ($1 \leq i \leq m$) is a polynomial over $\mathcal{A}$ in the variables $D^j x_i$ ($1 \leq i \leq m$, $j \leq n_i$). $D^j x$ is the j-th derivative of x: $D^0 x = x$; $D^{j+1} x = D(D^j x)$. Let $f(x)$

be a differential polynomial over $\mathcal{A}$ in one variable. Only $f(x)$'s having the following form will be considered:

$$(D^n x)^d + g_1(x)(D^n x)^{d-1} + \cdots + g_d(x),$$

where $g_i(x)$ $(1 \le i \le d)$ is a rational function over $\mathcal{A}$ in the variables $D^j x$ $(j < n)$. The order of $f(x)$, denoted by ord $f(x)$, is n; and the degree of $f(x)$ is d; $n \ge 0$ and $d > 0$. Let $\mathcal{B}$ be a differential field that extends $\mathcal{A}$. Suppose $b \in B - A$; b is a generic solution of $f(x)$ over $\mathcal{A}$ if b satisfies $f(x)$ but does not satisfy any $g(x)$ over $\mathcal{A}$ of lower order than $f(x)$. b is differential algebraic over $\mathcal{A}$ if b satisfies some $f(x)$ over $\mathcal{A}$; otherwise b is differential transcendental over $\mathcal{A}$.

The next proposition contains all the algebraic information needed to apply Blum's criterion (17.2) to DF_0.

Proposition 40.1 (A. Seidenberg [Sel]). *Let $\mathcal{A}$ be a differential field of characteristic* 0.

(i) *Suppose b is differential algebraic over $\mathcal{A}$. Then there exists an $f(x)$ such that b is a generic solution of $f(x)$ over $\mathcal{A}$, and such that all generic solutions of $f(x)$ over $\mathcal{A}$ are isomorphic over $\mathcal{A}$.*

(ii) *If $f(x)$ is a differential polynomial over $\mathcal{A}$, then there exists a b in some differential field extension of $\mathcal{A}$ such that b is a generic solution of $f(x)$ over $\mathcal{A}$.*

(iii) *$\mathcal{A}$ has a simple, differential transcendental extension. All such extensions are isomorphic over $\mathcal{A}$.*

Proof. (i) Let n be the maximum j such that $b, Db, \ldots, D^{j-1}b$ are algebraically independent over $\mathcal{A}$. Let $\mathcal{A}^*$ be the field (rather than the differential field) that results from adjoining $b, Db, \ldots, D^{n-1}b$ to $\mathcal{A}$. Then $D^n b$ is algebraic over $\mathcal{A}^*$, and the desired $f(x)$ is

$$(D^n x)^d + g_1(x, \ldots, D^{n-1}x)(D^n x)^{d-1} + \cdots + g_d(x, \ldots, D^{n-1}x),$$

where $f(x)$ is such that $f(b) = 0$ and

$$y^d + g_1(b, \ldots, D^{n-1}b)y^{d-1} + \cdots + g_d(b, \ldots, D^{n-1}b)$$

is irreducible over $\mathcal{A}^*$.

(ii) Suppose ord $f(x)$ is n. Let $b, b_1, \ldots, b_{n-1}$ be algebraically independent over $\mathcal{A}$. Define $D^j b = b_j$ for $0 < j < n$. Choose b_n so that

$$(b_n)^d + g_1(b, \ldots, b_{n-1})(b_n)^{d-1} + \cdots + g_d(b, \ldots, b_{n-1}) = 0.$$

Define $D^n b = b_n$. □

Blum's axioms for the theory of differentially closed fields (DCF$_0$) consist of DF$_0$ augmented by two schemas:

(1) if ord $f(x) >$ ord $g(x)$, then the system $(f(x) = 0, g(x) \neq 0)$ has a solution.

(2) if ord $f(x) = 0$, then $f(x)$ has a solution.

Clearly DCF$_0$ is a universal existential theory. Let $\mathcal{A}$ be a model of DCF$_0$, i.e. a differentially closed field of characteristic 0. By (2) $\mathcal{A}$ is algebraically closed. Suppose $f(x)$ is a differential polynomial over $\mathcal{A}$. By 40.1 (ii) $f(x)$ has a generic solution in some simple extension of $\mathcal{A}$. But by (1) $f(x)$ has "arbitrarily close" finite approximations of a generic solution in $\mathcal{A}$. If b is a generic solution of $f(x)$, then b satisfies the system $\{f(x) = 0\} \cup \{g(x) \neq 0 \mid \text{ord } g(x) < \text{ord } f(x)\}$. By (1) any system of the form $\{f(x) = 0\} \cup \{g_i(x) \neq 0 \mid i < n \ \& \ \text{ord } g_i(x) < \text{ord } f(x)\}$ has a solution in $\mathcal{A}$.

Theorem 40.2 (A. Robinson). DCF$_0$ *is the model completion of* DF$_0$.

Proof. (In the style of L. Blum.) It follows from 40.1 (ii) that every differential field of characteristic 0 has a differentially closed extension. By 17.2 it is enough to show every diagram of the following sort

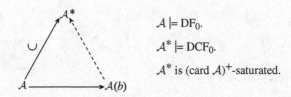

$\mathcal{A} \models \mathrm{DF}_0$.

$\mathcal{A}^* \models \mathrm{DCF}_0$.

$\mathcal{A}^*$ is (card $\mathcal{A}$)$^+$-saturated.

can be completed as shown. Suppose b is differential transcendental over $\mathcal{A}$. Let $\{g_i(x) \mid i \in I\}$ be any finite system of differential polynomials over $\mathcal{A}$. Since $\mathcal{A}^* \models \mathrm{DCF}_0$, there is a $c \in A^*$ such that $g_i(c) \neq 0$ for all $i \in I$. But then, since $\mathcal{A}^*$ is (card $\mathcal{A}$)$^+$-saturated, there is a $c^* \in A^*$ such that c^* is differential transcendental over $\mathcal{A}$. By 40.1 (iii) $\mathcal{A}(b)$ is isomorphic to $\mathcal{A}(c^*)$ over $\mathcal{A}$, so define $hb = c^*$.

Suppose b is differential algebraic over $\mathcal{A}$. Let $f(x)$ be the differential polynomial furnished by 40.1 (i). Thus b is a generic solution of $f(x)$ over $\mathcal{A}$, and all generic solutions of $f(x)$ over $\mathcal{A}$ are isomorphic over $\mathcal{A}$. Since $\mathcal{A}^* \models \mathrm{DCF}_0$, there are "arbitrarily close" finite approximations of a generic solution of $f(x)$ in $\mathcal{A}^*$. But then there must be a generic solution of $f(x)$ in $\mathcal{A}^*$, since $\mathcal{A}^*$ is (card $\mathcal{A}$)$^+$-saturated. □

Corollary 40.3 (A. Robinson). DCF_0 *admits elimination of quantifiers.*

Proof. By 13.2. □

Corollary 40.4 (A. Seidenberg). *Let $\mathcal{A}$ be a differentially closed field of characteristic 0, and let S be a finite system of differential polynomial equalities and inequalities in several variables over $\mathcal{A}$. If S has a solution in some differential field extending $\mathcal{A}$, then S has a solution in $\mathcal{A}$.*

Proof. Suppose S has a solution in $\mathcal{B} \supset \mathcal{A}$. By 40.1 (ii) there is no risk in assuming $\mathcal{B}$ is differentially closed. But then $\mathcal{B} \succ \mathcal{A}$ by 40.2. □

Corollary 40.3 implies the existence of an algorithm for deciding whether or not S, a finite system of differential polynomial equalities and inequalities in several variables over a differentially closed field $\mathcal{A}$, has a solution in $\mathcal{A}$. (Such an algorithm was first obtained by A. Seidenberg via elimination theory.) Suppose the members of $\mathcal{A}$ occurring in S are $a_1, \ldots, a_n$. The statement that S has a solution can be expressed by an existential sentence $F(\underline{a}_1, \ldots, \underline{a}_n)$ in the language of $T \cup D\mathcal{A}$. By 40.3 $F(\underline{a}_1, \ldots, \underline{a}_n)$ is equivalent to some quantifierless $G(\underline{a}_1, \ldots, \underline{a}_n)$. Thus S has a solution in $\mathcal{A}$ iff

$$\mathcal{A} \models G(\underline{a}_1, \ldots, \underline{a}_n).$$

Let $\phi(a_1, \ldots, a_n)$ be the least differential field of characteristic 0 containing $a_1, \ldots, a_n$. Computing the truthvalue of $G(\underline{a}_1, \ldots, \underline{a}_n)$ in $\mathcal{A}$ amounts to nothing more than evaluating finitely many differential polynomials over $\phi(a_1, \ldots, a_n)$ and noting which are zero and which are not.

Theorem 40.2 furnishes an effective method for transforming $F(\underline{a}_1, \ldots, \underline{a}_n)$ into an equivalent, quantifierless $G(\underline{a}_1, \ldots, \underline{a}_n)$. Let $\mathcal{Q}$ be the field of rational numbers with the trivial derivative, i.e. $Da = 0$ for all $a \in Q$. Then $\mathcal{Q} \subset \mathcal{A}$ for every differential field $\mathcal{A}$ of characteristic 0. By 40.2 $DCF_0 \cup D\mathcal{Q}$ is a complete theory. So by 40.3 there must exist a quantifierless $G(x_1, \ldots, x_n)$ such that

$$\mathrm{DCF}_0 \cup D\mathcal{Q} \vdash F(x_1, \ldots, x_n) \leftrightarrow G(x_1, \ldots, x_n).$$

$G(x_1, \ldots, x_n)$ can be found by recursively enumerating all proofs, since $\mathrm{DCF}_0 \cup D\mathcal{Q}$ has a recursive axiomatization.

Exercise 40.5 (I. Kaplansky). Let $\mathcal{A}, \mathcal{B}$ and $\mathcal{C}$ be differential fields of characteristic 0. Show there exists a differential field $\mathcal{D}$ so that the following diagram can be completed as shown.

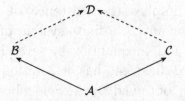

Exercise 40.6 (Carol Wood). The language of the theory of differential fields of characteristic p (DF_p) is that of DF_0 augmented by a 1-place function symbol $\frac{1}{p}$. The axioms for DF_p are: the axioms for fields of characteristic p; the two axioms for D included in DF_0; and

$$Dx = 0 \to (x^{\frac{1}{p}})^p = x$$
$$Dx \neq 0 \to x^{\frac{1}{p}} = 0.$$

(Caution: $x^{\frac{1}{p}}$ does not denote the usual p-th root of x whose p-th power must be x; the last axiom artificially fixes the function $x^{\frac{1}{p}}$ when $Dx \neq 0$; without it DF_p would lack a model completion.) Show DF_p has a model completion (known as the theory of differentially closed fields of characteristic p (DCF_p)). (cf. 41.2 and 41.8.)

Section 41

The Differential Closure

Morley's rank and degree machinery can be applied directly to the theory of differentially closed fields of characteristic 0 (DCF_0), because 40.2 implies DCF_0 is complete and substructure complete.

Proposition 41.1 (L. Blum). *Suppose $\mathcal{A} \models DF_0$, $T(x)$ is a differential polynomial over $\mathcal{A}$ of order n, and b is a generic solution of $f(x)$ over $\mathcal{A}$. Let $q_b \in S\mathcal{A}$ be the 1-type realized by b. Then:*

(i) *The Cantor–Bendixson rank of q_b is at most n.*
(ii) *If $\mathcal{A}$ contains solutions of $D^i x = j$ for all i and $j \geq 0$, and if $f(x)$ is $D^n x$, then the Cantor–Bendixson rank of q_b is n.*

Proof. Both (i) and (ii) are proved by induction on n.

(i) By 40.1 (i) there is an $h(x)$ of order n such that b is a generic solution of $h(x)$ over $\mathcal{A}$ and all generic solutions of $h(x)$ are isomorphic over $\mathcal{A}$. Suppose $p \in S\mathcal{A}$, $p \neq q_b$ and $h(x) = 0$ belongs to p. Then every realization of p is the generic solution of some differential polynomial of order less than n, and so by induction the Cantor–Bendixson rank of p is less then n. Consequently the Cantor–Bendison rank of q_b is at most n, since the formula $h(x) = 0$ defines an open

191

subset of $S\mathcal{A}$ whose only members are q_b and l-types of rank less than n.

(ii) The proof of 40.1 implies that q_b is completely specified by: $D^n x = 0$; $x, Dx, \ldots, D^{n-1}x$ are algebraically independent over $\mathcal{A}$. For each $m > 0$, let r_m be the l-type completely specified by: $D^{n-1}x = m$; $x, Dx, \ldots, D^{n-2}$ are algebraically independent over $\mathcal{A}$. Choose $a_m \in \mathcal{A}$ so that $D^{n-1}a_m = m$. Then r_m is completely specified by: $D^{n-1}(x - a_m) = 0$; $(x - a_m), D(x - a_m), \ldots, D^{n-2}(x - a_m)$ are algebraically independent over $\mathcal{A}$. By 28.3 r_m has the same Cantor–Bendixson rank as the l-type of a generic solution of $D^{n-1}x$; so by induction the Cantor–Bendixson rank of r_m is $n - 1$. Consequently the Cantor–Bendixson rank of q_b is n, since the formula $D^n x = 0$ defines an open subset of $S\mathcal{A}$ whose only members are q_b and l-types of rank less than n, and whose membership includes infinitely many l-types of rank $n - 1$. $\qquad\square$

Lemma 41.2 (L. Blum). DCF_0 *is totally transcendental. The Morley rank of* DCF_0 *is* $\omega + 1$.

Proof. It follows from 40.1 that DCF_0 is ω-stable. So by 31.6 DCF_0 is totally transcendental.

Let $\mathcal{A}$ be a universal domain for, and a model of, DCF_0. $\mathcal{A}$ exists by 16.4. It suffices, by 31.3, to show $\alpha_{\mathcal{A}}$, the Cantor–Bendixson rank of $S\mathcal{A}$, is $\omega + 1$. It follows from 41.1 (ii) that the Cantor–Bendixson rank of $S\mathcal{A}$ is at least ω; then by 41.1 (i), 40.1 (i) and 40.1 (iii), it is $\omega + 1$. $\qquad\square$

Let $\mathcal{A}$, $\mathcal{B}$ and $\mathcal{C}$ be differential fields of characteristic 0. $\mathcal{B}$ is prime differentially closed extension of $\mathcal{A}$ if the following diagram can be completed as shown whenever $\mathcal{C}$ is differentially closed.

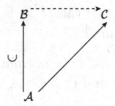

Theorem 41.3 (L. Blum). *Every differential field of characteristic 0 has a prime differentially closed extension.*

Proof. By 41.2 and 32.4. □

Theorem 41.4. *Let A be a differential field of characteristic 0. Any two prime differentially closed extensions of A are isomorphic over A.*

Proof. By 41.2 and 36.2. □

For each differential field A of characteristic 0, let $\bar{A}$ be the unique prime differentially closed extension of A afforded by 41.3 and 41.4. $\bar{A}$ is called the differential closure of A.

Suppose $B \supset A$. B is atomic over A if for each $n > 0$ and $\langle b_1, \ldots, b_n \rangle \in B^n$, there exists a finite system S of differential polynomial equalities and inequalities in n variables over A with the following properties:

(1) $\langle b_1, \ldots, b_n \rangle$ is a solution of S.
(2) All solutions of S are isomorphic over A.

Theorem 41.5. *$\bar{A}$ is atomic over A.*

Proof. By 41.2, 32.2 and 32.6. □

It follows from 41.5 that $\bar{A}$ is differential algebraic over A. Call $f(x)$ irreducible over A if all generic solutions of $f(x)$ over A are isomorphic over A. B is a normal extension of A if $A \subset B \subset \bar{A}$, and if for each differential polynomial $f(x)$ irreducible over A, either all or none of the generic solutions of $f(x)$ in $\bar{A}$

belong to $\mathcal{B}$. (It can happen that $f(x)$ is irreducible over $\mathcal{A}$ and has no generic solution in $\bar{\mathcal{A}}$.)

Theorem 41.6. *Let $\mathcal{B}$ and $\mathcal{C}$ be normal extensions of $\mathcal{A}$. Then any isomorphism of $\mathcal{B}$ and $\mathcal{C}$ over $\mathcal{A}$ can be extended to an automorphism of $\bar{\mathcal{A}}$.*

Proof. By 36.1, since 32.9 implies $\bar{\mathcal{A}}$ is atomic over $\mathcal{B}$ and over $\mathcal{C}$. $\qquad\qquad\qquad\qquad\qquad\qquad\qquad\qquad\qquad\square$

$\bar{\mathcal{A}}$ is minimal over $\mathcal{A}$ if there is no differentially closed $\mathcal{B}$ such that $\bar{\mathcal{A}} \supset \mathcal{B} \supset \mathcal{A}$ and $\bar{\mathcal{A}} \neq \mathcal{B}$. A case of maximal interest occurs when $\mathcal{A} = \mathcal{Q}$. $\bar{\mathcal{Q}}$ is not minimal over $\mathcal{Q}$, where $\mathcal{Q}$ is the field of rational numbers with the trivial derivative attached. L. Harrington has shown that $\bar{\mathcal{Q}}(= \langle \bar{Q}, +, \cdot, D \rangle)$ has a computable presentation; i.e. there is a one-one map from $\bar{Q}$ onto ω that transforms $+, \cdot$ and D into computable functions. Harrington's argument combines a Henkin style construction of $\bar{\mathcal{Q}}$ with the finite basis theorem for radical, differential polynomial ideals.

Exercise 41.7. Let $\mathcal{A}$ be a differential field of characteristic 0. Suppose $\mathcal{B}$ is a normal extension of $\mathcal{A}$ in $\bar{\mathcal{A}}$. Show $\bar{\mathcal{B}}$ is isomorphic to $\bar{\mathcal{A}}$ over $\mathcal{B}$.

Exercise 41.8. Show DCF_p (Exercise 40.5) is not totally transcendental.

Exercise 41.9. Show $S\bar{\mathcal{Q}}$ has isolated points of arbitrarily high finite rank.

Section 42

Some Other Reading

The most promising area of model theory neglected in this book is infinitary logic and in particular $\mathcal{L}_{\omega_1,\omega}$. The formulas of $\mathcal{L}_{\omega_1,\omega}$ are generated according to the rules of Sec. 4 and one further rule: if $\{F_i | i < \omega\}$ is a sequence of formulas, then $\&\{F_i \mid i < \omega\}$ is a formula. Although $\mathcal{L}_{\omega_1,\omega}$ allows countably infinite conjunctions and disjunctions, it forbids infinitely long quantifier prefixes. (It follows that some formulas lack prenex normal equivalents.) The axioms and rules of inference of $\mathcal{L}_{\omega_1,\omega}$ are those alluded to in Sec. 7 augmented by an ineluctable infinitary rule: if F_i is a consequence of S for each $i < \omega$, then $\&\{F_i | i < \omega\}$ is a consequence of S.

The completeness theorem for $\mathcal{L}_{\omega_1,\omega}$ reads: if a sentence F is consistent (under application of the axioms and rules of $\mathcal{L}_{\omega_1,\omega}$), then F has a model. The proof of completeness is a Henkin argument whose immediate ancestors are the proofs of 18.1 and 24.2. Most of the applications of $\mathcal{L}_{\omega_1,\omega}$ are animated by the absoluteness of the notion of consistency in the sense of $\mathcal{L}_{\omega_1,\omega}$. The best account of infinitary logic is given by Keisler [Kel].

Suppose $\mathcal{A}$ is a countable structure. Scott [Scl] showed there exists a sentence F of $\mathcal{L}_{\omega_1,\omega}$ (and of the same similarity type as $\mathcal{A}$) such that

$$\mathcal{B} \models F \quad \text{iff} \quad \mathcal{B} \approx \mathcal{A}$$

for every countable structure $\mathcal{B}$. One such F can be constructed by a canonical process whose unique end result is denoted by $F_\mathcal{A}$. The rank of $\mathcal{A}$ is defined to be the least number of steps needed to generate $F_\mathcal{A}$ from the atomic formulas. (It turns out that the rank of $\mathcal{A}$ is at most one more than the least ordinal not recursive in any real that encodes $\mathcal{A}$.)

A plethora of facts about stability and rank can be found in Shelah [Shl]. A complete countable theory T is stable if it is κ-stable for some infinite κ. T is superstable if it is κ-stable for all $\kappa \geq 2^\omega$. Shelah proved: if T is κ-stable for some κ such that $\kappa^\omega > \kappa$, then T is either superstable or totally transcendental. His proof combined the splitting device of 19.1 with a compactness trick. Lachlan [Lal] showed: if T is superstable and $n(T) > 1$, then $n(T) \geq \omega$. (His argument extended the dimensionality technique of Sec. 39.) A masterly use of rank occurs in Shelah [Sh2] where, for a theory T of arbitrary infinite cardinality κ, it is shown that T is categorical in every cardinality greater than κ if T is categorical in some cardinality greater than κ. The proof relies on some absoluteness facts to solve a type omitting problem via Chang's two-cardinal theorem.

Lachlan's inequality for $n(T)$, when T is superstable, is rendered comprehensible by a lemma of [La2]: if $\mathcal{A}$ is a model of T and $p \in S\mathcal{A}$ has a Morley rank, then the degree of p is 1. Baldwin's proof that the Morley rank of an ω_1-categorical theory is finite helps to make sense out of the theorems of Sec. 39.

$VF_{(p,0)}$ is the theory of p-valued fields of characteristic 0. A model of $VF_{(p,0)}$ consists of a field F of characteristic 0, a Z-group G (as in Sec. 17) with least positive element 1, a valuation ord: $F \to G$, a distinguished element p of F, and a cross section $X : G \to F$ such that ord $Xg = g$ and $X1 = p$; in addition the residue class field of F mod G is Z_p, the integers mod p. $H_{(p,0)}$, the theory of p-adic Hensel fields, is $VF_{(p,0)}$ coupled with Hensel's lemma. Clearly $VF_{(p,0)}$ can be presented as a universal theory. Ax and Kochen [Al] proved that $H_{(p,0)}$ is the model completion of $VF_{(p,0)}$. Their result can be obtained via Blum's criterion (Theorem 17.2) by following the general outline of the proof of Theorem 17.3. All the algebraic facts needed can be found in Kaplansky [Kal], as has been verified by Robinson [Ro2].

Suppose K is a field of characteristic 0. The ideas of the previous paragraph persist with little change when K replaces Z_p, i.e. when the residue class field is K rather than Z_p. The switch to K enabled Ax and Kochen [A2] to prove an asymptotic form of a conjecture of Artin, the most direct proof of which is in [Ro2]. Ersov [El] extended the above reasoning from Z-groups to arbitrary value groups.

Suppose $\mathcal{A}$ and $\mathcal{B}$ are elementarily equivalent structures. Keisler [Ke2] showed with the aid of the generalized continuum hypothesis (GCH) that there exist $\mathcal{A}^*$, an ultrapower of $\mathcal{A}$, and $\mathcal{B}^*$, an ultrapower of $\mathcal{B}$, such that $\mathcal{A}^* \approx \mathcal{B}^*$. Shelah developed another proof without GCH but less canonical than Keisler's: Keisler first constructs an ultrafilter D and then an isomorphism between $\mathcal{A}^I/D$ and $\mathcal{B}^I/D$; Shelah constructs the ultrafilter and the isomorphism simultaneously. The best reference for ultraproducts is [Ch1].

Shelah proved $n(\omega_\alpha, T)$, the number of models of T of cardinality ω_α, is at least card $(\alpha + 1)$, when T is totally transcendental, $n(\omega_1, T) > 1$ and $\alpha > 0$. (Keisler and Morley conjectured that $n(\omega_\alpha, T)$ is a nondecreasing function of α when $\alpha > 0$.) A short proof of Shelah's theorem has been discovered by Rosenthal [Rosl], who invokes GCH (to obtain some ultrafilters) and then circumvents it via absoluteness facts, the principal one being the absoluteness of the notion of total transcendentality.

References

[A1]   J. Ax and S. Kochen, Diophantine problems over local fields III, Ann. of Math. (1966), 83, 437–456.

[A2]   J. Ax and S. Kochen, Diophantine problems over local fields I and II, Amer. Jour. Math. (1965 and 1966), 87, 605–630 and 631–648.

[Ba1]   J. T. Baldwin, Ph.D. Thesis, Simon Fraser University (1971).

[Bar1] K. T. Barwise, Infinitary logic and admissible sets, Jour. Symb. Log. (1969), 34, 226–254.

[Bl1]   L. Blum, Ph.D. Thesis, Massachusetts Institute of Technology, 1968.

[Ch1]   C. C. Chang and H. J. Keisler, *Theory of Models*, North-Holland (1990), xvi+650pp.

[E1]   Y. Ershov, On the elementary theory of maximal normal fields, Algebra i Logica (1965), 31–70.

[Gö1]   K. Gődel, *The Consistency of the Generalized Continuum Hypothesis*, Princeton University Press (1940).

[Ka1]   I. Kaplansky, Maximal fields with valuations, Duke Math. Jour. (1942), 313–321.

[Ke1]   H. J. Keisler, *Model Theory for Infinitary Logic*, North-Holland (1971).

[Ke2]   H. J. Keisler, Ultraproducts and·elementary classes, Indag. Math. (1961), 23, 477–495.

[Ko1]   S. Kochen, Ultraproducts in the theory of models, Ann. of Math. (1962), 74, 221–261.

[La1]   A. H. Lachlan, On the number of countable models of a countable super-stable theory, in *Logic, Methodology and Philosophy of Science* IV, North-Holland (1973), 45–56.

[La2]   A. H. Lachlan, A property of stable theories, Fund. Math. (1972), 77, 9–20.

[Ma1]   W. Marsh, Ph.D. Thesis, Dartmouth College (1966).

[Mo1]   M. Morley, The number of countable models, Jour. Symb. Log. (1970), 35, 14–18.

[Pl1]   R. Platek, Ph.D. Thesis, Stanford University (1966).

[Ri1]   J. F. Ritt, Differential Algebra, Amer. Math. Soc. Publication (1950).

[Ro1]   A. Robinson, On the concept of a differentially closed field, Bull. Res. Council of Israel (1959), 113–128.

[Ro2] A. Robinson, Problems and methods of model theory, Centro Internazionale Matematico Estivo, Varenna (1968).

[Ros1] J. Rosenthal, A new proof of a theorem of Shelah, Jour. Symb. Log. (1972), 133–134.

[Sa1] G. E. Sacks, Effective bounds on Morley rank, Fund. Math. (1979), 103, 111–121.

[Sa2] G. E. Sacks, Bounds on weak scattering, Notre Dame Jour. of Formal Log. (2007), 48, 5–34.

[Sc1] D. Scott, Logic with denumerably long formulas and finite strings of quantifiers, *The Theory of Models*, North-Holland, 1965.

[Se1] A. Seidenberg, An elimination theory for differential algebra, Univ. of Calif. (1956), 31–66.

[Sh1] S. Shelah, Stability, the finite cover property and superstability, Ann. Math. Log (1971), 3, 271–362.

[Sh2] S. Shelah, Categoricity of uncontable theories, in Proc. of the Tarski Symposium, Proc. Symposium Pure Math XXV (1971), Univ. of Calif. Berkeley, 187–203.

[Va1] R. Vaught, Denumerable models of complete theories, *Infinitistic Methods*, Pergamon Press, 1961, 303–321.

Notation Index

Index